Contemporary Clinical Neuroscience

Series Editors
Ralph Lydic, Ph.D.
Department of Anesthesiology
The University of Michigan
Ann Arbor, Michigan
United States

Helen A Baghdoyan, Ph.D.
Department of Anesthesiology
The University of Michigan
Ann Arbor, Michigan
United States

For other titles published in this series, go to
www.springer.com/series/7678

David W. McCandless

Thiamine Deficiency and Associated Clinical Disorders

Humana Press

David W. McCandless
Department of Cell Biology & Anatomy
Rosalind Franklin University
3333 Green Bay Road
North Chicago, IL 60064
USA
david.mccandless@rosalindfranklin.edu

ISBN 978-1-60761-310-7 e-ISBN 978-1-60761-311-4
DOI 10.1007/978-1-60761-311-4

Library of Congress Control Number: 2009930939

Printed on acid-free paper

springer.com

This volume is dedicated to Sue, April, Elizabeth, and Andrew

Preface

The past 20 years have seen remarkable advances in neuroscience, neurology, imaging techniques, and diagnostic strategies. These advances have been successfully applied to many different diseases, including thiamine deficiency and associated clinical disorders. Syndromes such as beriberi, Wernicke's disease, Leigh's disease, African Seasonal Ataxia, and various inherited ataxias have all benefited from improved scientific approaches. It is timely therefore to examine new knowledge related to the effects of thiamine deficiency on organ systems and to specific thiamine-related clinical disorders.

The elucidation of the biochemistry of thiamine took many scientists many years to complete and is inextricably linked to the study of thiamine deficiency. Thiamine (vitamin B1) is a water-soluble compound which consists of a pyrimidine nucleus and a thiazole ring. A key derivative of thiamine is thiamine pyrophosphate (TPP), called cocarboxylase. This form is a coenzyme which participates in the decarboxylation of several essential intermediates involved in carbohydrate metabolism. ATP is involved in the transfer of phosphate to thiamine. The two key decarboxylation reactions are the decarboxylation of pyruvic acid and of alpha-ketoglutaric acid. The enzyme transketolase is another enzyme requiring TPP as coenzyme. This enzyme is important in the hexose monophosphate shunt. The dietary requirement for thiamine is based largely on the caloric intake, and primarily that of carbohydrates. Thiamine is widely distributed in plants and animal tissues and is found in high concentrations in yeast, beans, wheat germ, oats, ham, soy beans, etc.

Thiamine is a naturally occurring vitamin which has many functions in mammals. An early and frequently studied role of thiamine is its participation as coenzyme in enzymatic functions in energy metabolism. Clinical descriptions of thiamine-related disorders were published many years before the association of their effects to thiamine deficiency. For example, beriberi was first described in medical literature in 1642 by the Dutch physician Jacobus Bontius (some of his writings related to beriberi have been translated and appear in Appendix A). It was many years later before the association between thiamine and beriberi was discovered. The enormous interest in thiamine deficiency and the associated clinical disorders stems from the fact that these represent classical metabolic encephalopathies, and as such are reversible (treatable) when recognized early.

As stated above, thiamine plays an important role as coenzyme in reactions in the tricarboxylic acid cycle and in the hexose monophosphate shunt. Thiamine deficiency produces symptoms which are primarily localized to cardiac and nervous tissues. These symptoms can be rapidly reversed by thiamine administration, which led to the important concept of a biochemical lesion. The term "biochemical lesion" implies a lesion which occurs before morphological lesions, the presence of which may render the disorder less amenable to treatment and reversal.

Thiamine may also play a role in neurotransmission, as well as several other key biochemical pathways. Perturbation of these important metabolic events may also contribute to key neurochemical alterations, and their reversal in thiamine deficiency and associated clinical syndromes. The fact that frequently the symptoms associated with both pure thiamine deficiency and those in human disorders related to thiamine deficiency can be reversed (successfully treated) in early stages is promising. If the biochemical lesion can be recognized and corrected before the onset of less reversible structural changes, recovery and return to normal function is possible.

Clinical disorders shown to be directly related to thiamine deficiency include beriberi, Wernicke's disease, Leigh's disease, African Seasonal Ataxia (ASA), inherited ataxias, central pontine myelinolysis, and others.

Pure beriberi heart disease is rare in the United States; however, it is endemic in other parts of the world. The majority of cases of beriberi (85%) are subacute and mild. The remainder of cases may be more severe, and if not treated, death may ensue. Since beriberi is a complicated nutritional disorder, it is not necessarily a disorder of thiamine deficiency alone; however, treatment with thiamine usually reverses symptoms, sometimes in only hours. There is a long, rich history relating to beriberi, including the pioneering work of Christian Eijkman, which resulted in the awarding of a Nobel Prize (Dr. Eijkman's Nobel Prize speech is reproduced in Appendix C).

Wernicke's disease, first described in 1881 by Carl Wernicke, occurs frequently in the United States and is usually associated with chronic alcoholism. Like beriberi, Wernicke's disease is a complicated nutritional disorder, but in the early stages, many symptoms can be reversed by thiamine administration. Wernicke's disease may have a rapid onset, and neurological symptoms consist largely of confusion, ataxias, nystagmis, and ocular palsies. Other brain stem symptoms may include stupor and coma. Many Wernicke's patients may lapse into the more chronic disorder, Korsakoff's psychosis. In later stages, symptoms may be associated with structural alterations in specific brain sites. Classic studies by Dr. Maurice Victor and colleagues on Wernicke's disease were carried out over many years, and are the basis for many publications in this area.

Leigh's disease (subacute necrotizing encephalomyelopathy) is an uncommon disease with around 200 cases reported in the literature. This disorder seems to be genetically transmitted. In patients diagnosed with Leigh's disease, there was found in urine an inhibitory substance which blocked the formation of thiamine triphosphate (TTP). Low levels of TTP were found in tissues of Leigh's disease patients. This was postulated to be responsible for thiamine-deficiency-like symptoms seen in these patients. Leigh's disease usually begins to be symptomatic around postnatal month 6–12 following normal development. There appears lethargy, nystagmus, ataxia, failure to thrive, stupor, coma, and death. The lesions in the brain are usually more widespread than those in Wernicke's disease, but similar. Thiamine levels in Leigh's disease patients' blood and tissue are normal, and thiamine administration does not ameliorate the condition. A test for the inhibitor did not provide a conclusive indication of the presence of Leigh's disease, there being significant false positives and false negatives. Recent studies have shifted away from the inhibitor and examined supposed Leigh's disease patients for genetic defects including those with cytochrome c oxidase deficiency. Leigh's disease may represent one variation of the inherited ataxias. These studies and their ramifications will be described in depth. There are many unanswered questions relating to this fascinating disorder.

African Seasonal Ataxia (ASA) is another interesting recently described clinical entity which presumably has a thiamine-related foundation. This syndrome has been recently described in people who live in Western Nigeria and is characterized by ataxia, tremors, and decreased levels of consciousness. These symptoms occur during the rainy season of July through October. ASA usually follows a large carbohydrate meal. At its peak incidence, ASA can account for well over 70% of hospital and clinic admissions.

Various hypotheses have been advanced to explain ASA; however, strong evidence supports a mechanism related to thiamine deficiency. There is a clinical triad of cerebellar ataxia, ocular disturbances, and encephalopathy usually seen in acute thiamine deficiency. Upon examination of the dietary intake of the low socioeconomic strata of patients, it was found that almost all had consumed significant amounts of roasted silkworm larvae *Anaphe venata*. The availability of these larvae in the marketplace corresponds to the rainy season. The larvae represent a valuable protein source for rainforest people.

The practice of entomophagy in low socioeconomic cultures is accepted. Protein sources are relatively scarce for these people, who subsist largely on carbohydrate-rich diets. Subsequently, it has been shown that there is a thiaminase present in the *Anaphe venata* larvae. During the rainy season, these larvae fall from specific trees, and are gathered and sold in markets. Subsequent consumption of larvae containing a thiaminase by people ordinarily eating carbohydrate-rich diets can explain the rapid onset of symptoms resembling those of thiamine deficiency. As a corollary to this recent description are earlier descriptions of similar outbreaks of thiamine deficiency. There was, for example, the outbreak of the so-called Chastek paralysis, a disorder of silver foxes on a fox ranch in Minnesota. These foxes were fed raw fish and within a few weeks developed ataxia, changes in consciousness, seizures, and death. Pathologically, brain lesions resembled those seen in thiamine deficiency. Subsequent work showed the presence of a thiaminase in the viscera of the raw fish, which had precipitated the disorder.

Inherited ataxias are a group of relatively rare neurological disorders genetically transmitted, which have as a common denominator ataxia and the possibility of successful thiamine treatment. These diverse, yet related ataxias include Refsum's disease and Friedreich's ataxia. This group of disorders has a defect in the enzyme pyruvate decarboxylase. Pathologically, these disorders show mitochondrial damage in selective brain regions. Treatment regimes include thiamine and ketonic diet therapies.

The disorder called central pontine myelinolysis (CPM) represents one variant almost always associated with some other serious chronic disease. Over one half of the reported cases were associated with alcoholism and Wernicke's disease. As the name implies, the key feature is a pathological lesion consisting of a symmetric focus of demyelination localized at the center of the pons. Microscopically, there is demyelination of medulated fibers. A variation of CPM are cases which, in addition to the lesion in the pons, demonstrate demyelinating lesions in other brain regions including the thalamus, cerebellum, cerebral cortex amygdala, putamen, etc. In spite of the predominance of alcoholism and Wernicke's disease as a feature of CPM patients, there are many instances of CPM in serious disorders not associated with thiamine involvement. The relation between thiamine and CPM therefore remains somewhat unclear.

Marchiafava-Bignami disease is yet another relatively rare disorder, which has been associated with thiamine deficiency. This clinical entity has as its distinguishing characteristic, lesions in the corpus callosum. These lesions lead to specific signs and symptoms, which together with MRI can lead to early diagnosis. Before imaging advances, the diagnosis was almost always made at autopsy. As a result of rapid diagnosis, treatment regimes involving thiamine administration have evolved, and cases of Marchiafava–Bignami disease have been successfully treated.

In summary, thiamine deficiency and associated clinical disorders represent an intriguing area of both basic and clinical investigation. Modern imaging and diagnostic strategies have facilitated the rapid treatment, and potential reversal of these clinical disorders. The fusion of laboratory and clinical knowledge serve as an example of how research can translate to successful treatment. This book is designed to bring together cogent results from both basic and clinical investigation. These data will be of interest to neurologists, internists, nutritionists, biochemists, neurochemists, neuroscientists, and many others with interest in thiamine deficiency. Many questions regarding these clinical disorders as well as thiamine deficiency remain unanswered. We hope this book may serve to stimulate further investigations in these areas.

Acknowledgments

The author wishes to thank Dr. Joseph X. DiMario for support and consultation regarding several aspects of this book. The author also thanks Dr. Roland Auer and Dr. Jeffrey Joseph, University of Calgary, Department of Pathology and Laboratory Medicine, Calgary, Alberta, Canada, and Dr. Michael D. Norenberg, Director of Neuropathology, Department of Pathology, University of Miami School of Medicine, Miami, FL, for supplying many fine figures from brain slices, and photomicrographs. These illustrations add much to the visualization of features such as the highly localized nature of lesions in areas such as the mammillary bodies. Dr. Bola Adamolekun, Department of Neurology, University of Tennessee College of Medicine, graciously supplied me with several pounds of *Anaphe venata* larvae, shown in Figs. 1 and 2 of the chapter "African Seasonal Ataxia."

This volume would not have been possible were it not for the dedicated participation of Ms. Cristina Gonzalez. She was tireless in searching out old and new references, preparing tables, and organizing figures and figure permissions. Ms. Vilmary Friederichs also participated in organizing the manuscript and gave advice in the preparation of the volume. My son Dr. Jeffrey McCandless, NASA Ames Research Center, Palo Alto, CA, helped in the design of many excellent figures and in improving others. Finally, Ms. Ann Avouris, editor at Springer Science and Business Media, was outstanding in answering questions, helping obtain permissions for use of previously published figures, and in general shepherding this manuscript through the publishing process.

Contents

Early Chemistry

The early chemistry of thiamine deficiency goes back many years. The purpose of this introductory chapter is to review some of the remarkable discoveries relating to thiamine, thiamine-dependent enzymes, the relation to a deficiency of thiamine, and results, which led to the awarding of a Nobel prize to Christiaan Eijkman in 1929 in physiology or medicine for his work involving thiamine.

Remember that the first medical description of beriberi was published in 1642 by Jacobus Bontius (see Appendix A for a translation). Christiaan Eijkman, a physician, was able to demonstrate 250 years later that birds that were fed a diet of polished rice gradually progressed to a point where they became moribund and died. Eijkman had earlier tried unsuccessfully to isolate infectious agents that might have been responsible for beriberi. Eijkman next demonstrated that if the birds were fed unpolished rice, their symptoms reversed. The reversal feature of what was thiamine deficiency (but not known by Eijkman) was demonstrated very early in the investigations of the neurological (polyneuritis) outcomes of feeding polished rice. Also, Eijkman observed that the symptoms seen in pigeons closely resembled those seen in human beriberi.

The following is a description from Eijkman quoted verbatim concerning the symptoms in birds:

> "The beginning of the disease is characterized by an unsteady gait which first of all manifests itself in walking about on the perch, as if the animal cannot squeeze its toes around it firmly enough, and must exert itself in order not to fall off. The disturbance in mobility soon increases in intensity and speed. The fowl no longer has the strength to climb up; because of weakness it holds its limbs spread apart and bent at the knee and ankle joints; when running it frequently collapses or falls over. Finally, it remains lying on its side and in its fruitless efforts the developing paralysis of the wing muscles also becomes noticeable. The paralysis of the body musculature rapidly progresses from below upward."
>
> "The involvement of the peripheral nerves is the most important feature that post-mortem investigations reveal to date. It involves both the sensory and motor portions, which occur focally in the nerve trunks, and produces the picture of non-inflammatory atrophic degeneration such as is observed after transection of a nerve in the distal portion of the divided fragment. However, definite changes in the spinal cord and spinal cord roots are also not lacking. These show, likewise, the appearance of degeneration and atrophy."

D.W. McCandless, *Thiamine Deficiency and Associated Clinical Disorders*,
Contemporary Clinical Neuroscience, DOI 10.1007/978-1-60761-311-4_1,

Some years later, Eijkman and Vorderman (Eijkman, C. and Vorderman, A., 1897) were able to show that it was the "polishing" of the rice that was causing the symptoms of polyneuritis that were prevalent in birds and in man suffering the disease beriberi. Later it was discovered that a very small amount of a substance called "vitamine" by Funk (1911) was the compound which was essential for normal health, and the substance whose absence was responsible for symptoms in birds and man. The final "e" of vitamine was dropped when it was learned that the substances were not amines.

Early on Eijkman knew that polyneuritis was linked to the type of rice eaten. He believed that the symptoms were due to an infectious agent or toxin on the rice. Later it was shown that there was no toxin or infectious agent associated with the rice or its preparation.

Many investigators tried unsuccessfully to purify and isolate the mysterious vitamin until Jansen and Donath (1926) succeeded. It was a decade later before the vitamin was synthesized in the laboratory of R. R. Williams. Jensen and Donath sent a small amount of the compound to Eijkman who was able to show that as little as 2 μg was enough to protect pigeons from the deleterious effects of a diet of polished rice. Dr. Christiaan Eijkman received the Nobel Prize in Physiology or Medicine in 1929. His acceptance speech appears in Appendix C.[1]

Later in 1936, the structure (Williams, R., 1936), then the synthesis of the vitamin (Williams, R. and Cline, J., 1936) appeared as a communication to the editor of the Am. J. Chem. Soc. "Many laboratories were working on this problem, and later were able to confirm Williams" work.

The structure for thiamine put forth by Williams is as follows (Fig. 1):

Fig. 1

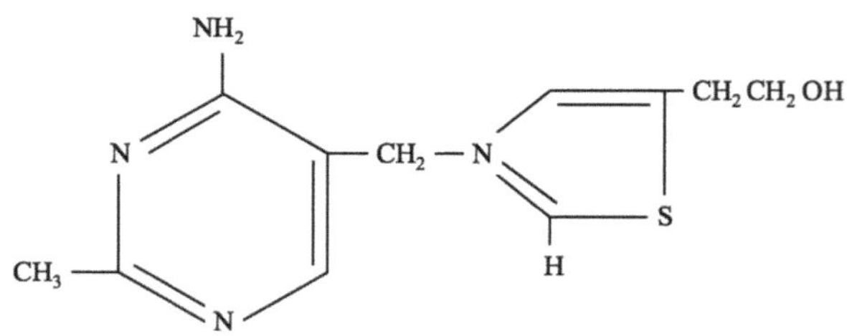

Thiamine

Using liquid ammonia, Williams was able to produce a free base of thiamine, $C_6H_{10}N_4$. A couple of months later, the synthesis of thiamine was briefly outlined by Williams. The synthesis included ammonia, $POCl_2$, and HBr. The structure and synthesis of thiamine was the culmination of many years of hard work.

[1] It is important to note that Frederick Gowland Hopkins shared the Nobel Prize in Physiology or Medicine with Eijkman in 1929. It was awarded to Hopkins for his work on vitamins and for his finding that muscle contraction leads to lactate accumulation.

The active form of thiamine is thiamine pyrophosphate or cocarboxylase. This active form has a structure as follows (Fig. 2):

Fig. 2

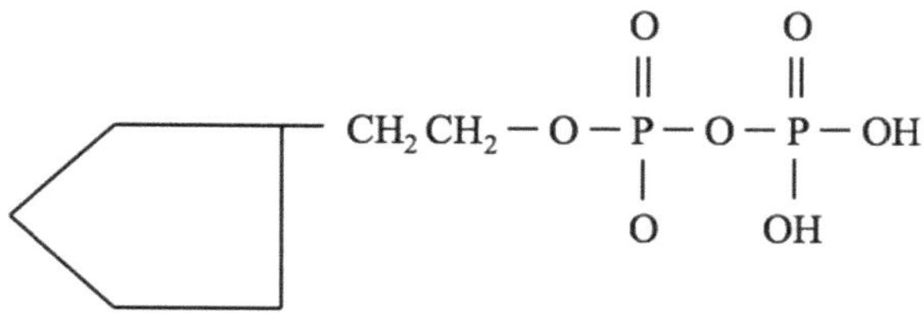

Thiamine pyrophosphate (cocarboxylase)

One of the first to synthesize thiamine pyrophosphate was Weil-Malherbe (1940), who synthesized the compound by treating thiamine bromide with silver pyrophosphate. It was also shown about the same time that yeast, bacteria, and animal tissues could also generate thiamine pyrophosphate enzymatically (Fig. 3).

Fig. 3

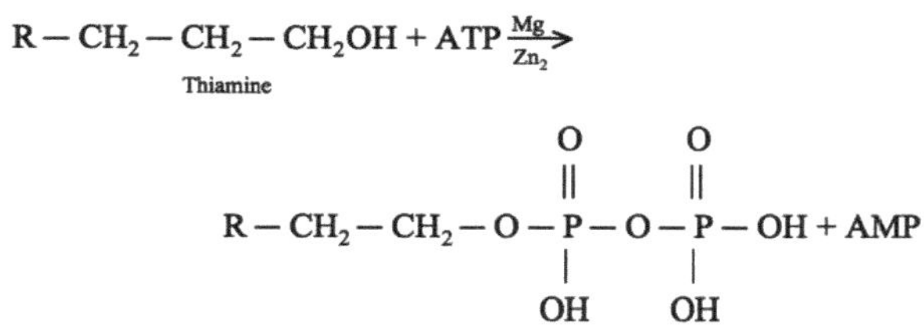

Thiamine pyrophosphate (cocarboxylase)

Finally it was shown (Lohmann and Schuster, 1937) that thiamine pyrophosphate was the coenzyme "active" form of thiamine.

Three key enzymes in carbohydrate metabolism require thiamine in its phosphorylated form (thiamine pyrophosphate) in order to function properly. The two enzymes pyruvate and alpha-ketoglutarate dehydrogenases are of particular interest because of their pivotal role in metabolism. Pyruvate dehydrogenase is important in lipid synthesis, gluconeogenesis, and oxidation via the Krebs (TCA) cycle. The Krebs cycle produces much needed energy in the form of ATP. The third thiamine-requiring enzyme is transketolase, which is the rate limiting step in the hexose monophosphate shunt, which produces reducing power (NAD/NADH) and ribose.

The pyruvate dehydrogenase complex is regulated largely by product inhibition. ATP in micromolar concentrations is able to inhibit kidney pyruvate dehydrogenase (Reed, L., 1976). Similar compounds such as ADP, CTP, and GTP were not shown to be regulatory. Magnesium is also an important component of the enzyme complex, and when pyruvate dehydrogenase is phosphorylated, but inactivated, Mg can activate the enzyme.

The alpha-ketoglutarate dehydrogenase complex is large (2.7 million M.W.). It can be separated from other dehydrogenase complexes by fractionation (Koike, M. et al., 1976). Highly purified enzymes were analyzed for substrate specificity. Results showed that the pyruvate dehydrogenase enzyme complex has a rather wide range of substrate specificity in that it can decarboxylate other substrates besides pyruvate, including alpha-ketobutyrate, alpha-ketovalarate, and alpha-ketoceproate, among others. The enzyme alpha-ketoglutarate dehydrogenase was found to be specific for alpha-ketoglutarate. Both enzymes catalyze the first step of the two enzymes which decarboxylate pyruvate or alpha-ketoglutarate.

The possibility of important regulation of the pyruvate dehydrogenase enzyme is based on its critically important location in glycolysis, in lipid metabolism, and in the Krebs cycle. Studies have shown that the enzyme is inhibited by its products such as acetyl-CoA and NADH, and the enzyme reverts to normal activity in the presence of NAD and CoA. Other keto acids mentioned above were inhibitory for the pyruvate dehydrogenase enzyme complex. The binding of thiamine pyrophosphate to pyruvate dehydrogenase is loose, making the activity highly dependent on available thiamine pyrophosphate.

Examination of the effect of the three major methods of producing thiamine deficiency in mammals (diet only, diet plus oxythiamine, and diet plus pyrithiamine) shed some light on the enzymes that thiamine pyrophosphate serves as coenzyme (Gubler, C. and Johnson, L., 1967). Animals were Sprague-Dawley rats, and controls received 10 μg/100 g/day of thiamine supplementation. After sacrifice, tissue was separated, and pyruvate decarboxylase and alpha-ketoglutarate decarboxylase was measured in heart, liver, kidney, and brain. Results showed that pyrithiamine plus deficient diet decreased pyruvate decarboxylase in all tissues examined, while alpha ketoglutarate decarboxylase was decreased only in brain and kidney. Oxithiamine plus deficient diet caused a decrease in pyruvate decarboxylase in liver, heart, and kidney, while alpha-ketoglutarate was only reduced in the heart. Replacement of the carbohydrate diet (thiamine deficient) with a high-fat diet (which could have "fed" the system) had only a trivial effect.

Further studies showed that pyrithiamine was an inhibitor of the enzyme thiamine pyrophosphokinase, thus blocking the formation of thiamine pyrophosphate. Oxythiamine formed oxythiamine pyrophosphate, which was in turn an inhibitor of both pyruvate decarboxylase and alpha-ketoglutarate decarboxylase, as well as transketolase.

The structure of the actual enzymes pyruvate dehydrogenase and alpha-ketoglutarate dehydrogenase have been studied in part with electron microscopy (Reed, L., 1967).

The thiamine-requiring enzyme alpha-ketoglutarate dehydrogenase is actually composed of three enzymatic subunits: alpha-ketoglutarate decarboxylase, dihydrolipoyl transsuccinylase, and dihydrolipoyl dehydrogenase. When the three subunits are combined, they automatically form a large molecule, which resembles the structure of the alpha-ketoglutarate dehydrogenase.

The thiamine-requiring enzyme pyruvate dehydrogenase is also composed of three subunits: pyruvate decarboxylase, dihydrolipoyl transacetylase, and dihydrolipoyl dehydrogenase. This composition is very similar to that of alpha-ketoglutarate dehydrogenase. These three components of pyruvate dehydrogenase form a three-dimensional configuration such that during enzyme functioning, the components are brought close enough to each other to function properly.

Electron microscopic studies of pyruvate dehydrogenase enzyme show a polyhedral form with a diameter of about 400 Å. Electron micrographs show that the transketolase moiety is located in the center of the polyhedron, and the other two complexes are located around the transacetylase. Electron micrographs also show the transacetylase moiety has eight subunits at the eight points of a cube. The transacetylase component is key to the overall three-dimensional form of the pyruvate dehydrogenase complex.

In terms of the biosynthesis of pyruvate dehydrogenase, studies indicate that the three-dimensional configuration of the complex is under direct genetic control. Little wonder that such a complex structural component, under genetic control, might have a variety of point mutations, any one of which could affect the functional capacity of either of these two thiamine-requiring enzyme complexes, pyruvate dehydrogenase and alpha-ketoglutarate dehydrogenase.

Many of the structural/three-dimensional studies have initially been performed in bacteria. Later studies in mammalian tissues such as heart and kidney show quite similar structural results. Electron microscopic examination show striking similarities to the enzyme in bacteria-based studies. Even down to basic detail, such as the position of the subunits on the transacetylase, components are similar to the bacteria counterpart.

A specific three-dimensional structure may enhance the ability of the complex to perform its duty. In addition, there may be a significant increase in the efficiency of the complex to dehydrogenate substrate. The close association of the three components should facilitate activity. This might be a sort of evolution in terms of efficiency of enzyme capacity to function.

Experimental evidence supports the concept that there is a protein which binds thiamine in bacteria (*Escherichia coli*) (Nose, Y. et al., 1976). Data examining the inhibition of thiamine binding to protein showed that classical thiamine analogs such as oxythiamine and pyrithiamine had little effect on binding, but that thiamine monophosphate, thiamine diphosphate, and thiamine triphosphate had over 50% inhibition. Chemical treatments such as iodine and bromosuccinimide had an over 50% inactivation of the binding of thiamine to protein.

The regulation of pyruvate dehydrogenase has been examined (Shen, L. et al., 1968). In this study, pyruvate dehydrogenase was isolated from *E. coli*. The enzyme was purified by protamine sulfate precipitation and elution (Koike, M. et al., 1976). Results showed that pyruvate dehydrogenase is inhibited by S-acetyl-CoA, and this inhibition is directed to the first reaction enzyme, a thiamine-requiring enzyme. Further regulating this first enzyme are nucleotide concentrations. Thus AMP, ADP, and ATP act to increase the rate of activity of the enzymes whether S-acetyl-CoA is present or not. While ATP in this preparation was stimulatory, in vivo it is probably inhibitory. It is important to consider only real physiological variables in vivo in attempting to assess regulation of pyruvate dehydrogenase.

Since the brain is not a gluconeogenic organ, the regulation of the thiamine-dependent enzyme pyruvate carboxylase by rat brain mitochondria has been investigated (Patel, M. and Tilghman, S., 1973). Rat brain mitochondria were isolated after decapitation and viability assured by measuring the P:O ratio and respiratory quotient. When all cofactors were present, the mitochondria fixed $H_{14}Co_3$ at a rate of 269 nmoles/30 min/mg protein. If pyruvate, $MgCl_2$, or ATP were left out of the mixture, fixation was nearly zero. Omission of Pi reduced decarboxylation by 65%. The combination of adenonucleotide and Pi data tend to show that the decarboxylation is regulated by intramitochondrial adenine nucleotide concentration. Except for citrate and isocitrate, the other dicarboxylate intermediates of the TCA cycle tended to decrease the decarboxylation of pyruvate as much as 70%.

The concentration of pyruvate also played a role in the regulation of mitochondrial decarboxylation. The role in regulation for the cofactor acetyl-CoA was not examined, but speculation is that it plays a role. The acetyl moiety of acetyl-CoA is vital for lipogenesis as well as the formation of acetylcholine. These are cytosolic processes, and the acetyl-CoA does not cross mitochondrial membranes. Data tend to show that citrate may act as a transporter to carry the acetyl moiety to the cytoplasm.

In another study (Tomita, I. et al., 1974), analogues of thiamine pyrophosphate were examined for their effects on pyruvate decarboxylase and transketolase, both thiamine-requiring enzymes. Results showed that both pyruvate decarboxylase and transketolase reconstitution was inhibited by thiazole pyrophosphate as well as by methyl and benzyl thiazole pyrophosphate.

The phosphorylated esters of 2-methyl and 2-benzyl thiazoles were inactive in combination with apopyruvate decarboxylase, or with apotransketolase. It appears to be the pyrimidine moiety of thiamine pyrophosphate which is important in binding studies, causing hydrogen bonding to take place. It is speculated that the pyrimidine moiety could act to tighten the binding of coenzyme by negating ionic repulsion.

Another study (Taylor, S. et al., 1975) describes the effects of the ATP/ADP ratios and other variables on the activity of pyruvate dehydrogenase. A rapid freezing method was developed to minimize any possible artifact change between active and inactive pyruvate dehydrogenase.

Results showed that the rapid freezing method was successful in extracting pyruvate dehydrogenase with little interconversion between the active and inactive forms of the enzyme. The investigators found a good correlation between the active form of pyruvate dehydrogenase and the intramitochondrial ATP/ADP ratio. Also noted was that pyruvate increased the active form of pyruvate dehydrogenase without any alteration in the ATP/ADP ratio. Octanoate acted to lower the active form of the enzyme, also without any significant change in the ATP/ADP ratio.

These data, taken together, lend strong support for the concept that the ATP/ADP ratio is an essential component for pyruvate dehydrogenase activity. This observation might be best explained by the fact that ADP is a competitive inhibitor of active pyruvate dehydrogenase kinase. The authors speculate that the ATP/ADP ratio in mitochondria may also be a key regulator of the active form of the enzyme in intact cells.

The developmental pattern of pyruvate carboxylase has been studied in newborn rats because of its importance in gluconeogenesis (Chang, L., 1977). In these studies, animals from –5 days (premature) to 36 days after birth were utilized. Results showed that about 70% of pyruvate carboxylase activity was located in the mitochondrial fractions of liver. Twenty percent was found in the nuclear fraction. Only about 5% of adult levels of pyruvate carboxylase were found in fetal liver. The levels of enzyme remained low until birth, when activity levels rapidly rose, reaching a peak at about 5 days after birth. The actual rise after birth occurred after the first 8 hours, suggesting that the birth process in and of itself was not significant.

The developmental pattern of pyruvate carboxylase in vitro agrees with that seen previously in vivo (Snell, K., 1974). The author speculates that the rise in activity of the enzyme could be related to liver maturation. More likely is that the rise after birth is due to a sudden change from a high-carbohydrate diet (transplacental) to a diet of milk high in fat (nursing). This would spell a clear adaptation to changing dietary balance.

Early Thiamine Deficiency

Early thiamine deficiency studies were performed on pigeons because these birds were uniquely sensitive to polished rice. Symptoms frequently were evident by 21 days, which was as early as or earlier than any other animal model. This sensitivity also took the form of pronounced symptoms and certain death when not reversed with thiamine. The reversal feature was key in determining the significance of biochemical changes. If symptoms reversed with thiamine treatment along with key substrates such as decreased lactate, then this implied a role for lactate in the generation of the symptoms (Fig. 1).

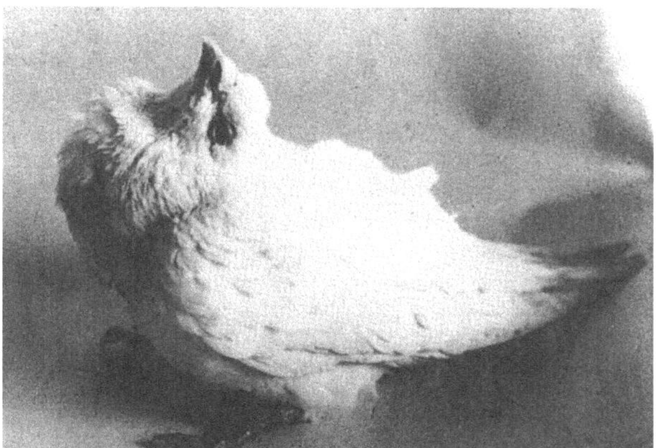

Fig. 1 Pigeon showing opisthotonic posturing due to thiamine deficiency. Reproduced from Yoshinori (1995) with permission from Springer

Several studies from about 1910 to 1916 were pivotal in developing concepts relating to vitamin physiology and biochemistry. For example, it was discovered that certain chemicals were necessary for continued health, and these were termed "vital amines," from which the word vitamin was coined. It was shown that adding carbohydrate to a vitamin-free diet brought about the development of symptoms more rapidly. This suggested that the carbohydrates require vitamins, and the addition of carbohydrates increases vitamin demand and consumption.

D.W. McCandless, *Thiamine Deficiency and Associated Clinical Disorders*, Contemporary Clinical Neuroscience, DOI 10.1007/978-1-60761-311-4_2,

Casimir Funk was instrumental in some of these early studies. In one report (Funk, C., 1911), he placed pigeons on several diets and the results were compared. The data showed that all pigeons placed on diets with high concentrations of carbohydrates developed hyperglycemia. Pigeons maintained on a vitamin-free diet also developed hyperglycemia. When hyperglycemic animals were treated with vitamins, the hyperglycemia diminished and liver glycogen levels increased. These studies show early correlations between carbohydrate metabolism and vitamins.

By the mid-1920 s several studies had been published, sometimes with conflicting data. Most investigators, however, were in agreement that animals deprived of vitamin B had significantly decreased oxygen consumption in various tissues. This was largely attributed to a reduction in the amount of activity of oxidizing enzymes. However, even in related measurements such as the respiratory quotient, there was disagreement. Hess and Messerle (1922) suggested that there was a similarity between the polished rice-fed beriberi pigeons and the result of sublethal doses of cyanide. He also speculated that the biochemical results in animals eating polished rice were manifest in a decrease in the amount or efficiency of oxidizing enzymes.

Given the somewhat controversial situation regarding various hypotheses that had been put forth to explain the relation of these findings to vitamin B metabolism, a rigorous study was published (Drummond, J. and Marrian, G., 1927). In one set of experiments, the authors compared the rates of reduction of methylene blue as a hydrogen acceptor to assess the oxidation/reduction status. Normal controls and beriberi tissues (pigeon breast muscle) were compared. Results did not confirm studies by Hess (Hess and Messerle, 1922). The Hess results showed that it took longer for the vitamin-deficient (beriberi) tissue to produce a reduction of methylene blue than in the case of normal non-deficient tissue. In the study by Drummond and Marrian, there was a small increase in time for the reduction of methylene blue, but it was not deemed significant.

Similarly when these investigators examined oxygen consumption in normal and beriberi tissues, results showed no real difference in breast muscle and only a small decrease in liver. These workers also documented the weight loss seen in the final days on vitamin B deficient diet. The authors conclude that on the basis of their findings, there is no evidence to support the view that vitamin B is essential for the functioning of the oxidative mechanism in tissues. The authors do point out that they cannot rule out the possibility that some small stimulation in oxidation might occur in intracellular locations. This could explain the small rise in temperature seen when animals deficient in vitamin B are administered extracts containing the vitamin.

In another study (Kinnersley, H. and Peters, R., 1930) published 2 years later, the relation between lactic acid in the brains of avitaminosis pigeons and normal control birds was described. Pigeons were stunned and the head was cut off and quickly frozen in liquid air. Frozen brain samples were extracted in 10% TCA, and following suitable preparation, lactic acid in the brain and in the blood was measured in normal pigeons and in opisthotonic pigeons. The vitamin-deficient pigeons were made opisthotonic by feeding them with polished rice. Results showed a much higher concentration of lactic acid in the brains of opisthotonic pigeons than in that of normal control birds.

Worried that some artifact of lactic acid might have occurred due to the few seconds delay before the brain was frozen, the authors constructed a special guillotine. This devise was described as "T" shaped with two blades at different levels. The lower blade decapitated the pigeon, while the second, higher blade split the skull. This effectively severs the blood supply to the brain, and the splitting of the skull facilitates more rapid freezing.

Results using the improved guillotine showed elevated lactic acid, as was seen earlier. The increased lactic acid was noted during the opisthotonic stage, not earlier, and the elevated lactic acid rapidly returned to normal when the animals were treated (reversed). Results also showed that stunning produced higher levels of lactic acid than did anesthesia with ether. The authors conclude that lactic acid levels seen in brain are more than those that could result from blood levels, thereby suggesting storage of carbohydrate in brain. The rapid reversal with vitamin B1 indicates no other food material or ions could have been deficient enough to produce the severe symptoms seen in these pigeons.

It is amazing to see that in 1925 investigators (Kinnersley and Peters) knew about the lability of brain metabolites. They designed and built a two-bladed guillotine to first cut off the head and then split the skull, rendering easier access to the brain for freezing. This worked to prevent or minimize artifacts after death. Interestingly, in the late 1960 s into the early 1970 s there was intense concern about the very real risk of changes occurring in brains of experimental animals following killing. In fact, devices were designed to minimize changes. These included freeze blowers, microwave ovens, and brain choppers which "fixed" or froze tissue in milliseconds, rendering the measurement of "true" levels of labile metabolites. One wonders if those investigators knew of Peters' attempts along the same lines, which were done 50 years earlier.

In another study by Peters (Gavrilescu, N. and Peters, R., 1931), oxygen uptake was determined in pigeon brain regions during vitamin B deficiency induced symptoms. In this study, oxygen uptake was performed rapidly after killing, the elapsed time being 12–17 minutes. The guillotine developed and described above was used. Brain regions were isolated and measured separately. Regions sampled included cerebellum, cerebrum, optic lobes, and "lower parts." The speed of tissue preparation was attributed in part to a new air-damped Sartorius balance. The Homer pigeons were fed in the standard way for producing severe symptoms.

Results showed the usual symptomology of opisthotonus, and death if not treated. Oxygen uptake was determined in these birds with head retraction or just before these symptoms were apparent. Oxygen uptake in symptomatic birds' cerebellum was comparable to normal birds, but values in the cerebrum, optic lobes, and lower parts were decreased in symptomatic as compared to normal birds. The most significant decreased areas were the cerebrum and optic lobes. It was further shown that the oxygen uptake reverted to normal when the animals were "cured."

The authors state that these oxygen uptake data, together with earlier published lactate data, show conclusively that the disorder in vitamin-deficient pigeons represents a biochemical defect that is central in origin, not peripheral. The rapid reversal of opisthotonus and decreased oxygen uptake with vitamin B treatment prove that these features are due to vitamin B1 deficiency, not starvation. The fact that the cerebellum was unaltered as regards oxygen uptake serves as an internal control, arguing against changes in blood circulation in the affected pigeon brain. The authors state that they do not think some circulating toxin is involved.

It is interesting to note that even today (2009) many neurochemical studies are performed on whole brain, neglecting the brain's heterogeneity, whereas 80 years ago regional brain studies were being performed, and regional differences in neurochemical responses to perturbations were being described.

In a later study (Gavrilescu, N. and Peters, R., 1931) the issue of the possible local lack of vitamin B1 was addressed. Arguing in favor of a local change are data showing regional alterations in biochemical paradigms such as oxygen uptake, and the rapid (30 minutes) reversal of symptoms by injection of thiamine into the cranium. The counterargument was that some hormone or some other substance might be stimulated and/or released into the circulation by vitamin B1, thereby causing the reversal of biochemical changes and symptoms. The authors chose to examine oxygen uptake in vitro when vitamin was added to a brain preparation from symptomatic pigeons.

Animals used were pigeons in the final stages of vitamin B1 deficiency and showing the symptom of opisthotonus. Brain areas used were optic lobes, cerebrum, and lower brain parts. Tissue was collected as before, minced, and oxygen uptake measured as previously described. After equilibration, vitamin B1 was added to the mixture. Results showed that when B1 was added to optic lobe and to the lower parts, there was a clear increase in oxygen uptake by the homogenate. Various methods of inactivation of vitamin B1 were tried, and only when the vitamin was completely inactivated did the addition fail to increase oxygen uptake.

Earlier results indicated the cerebrum acted differently, and in these experiments the effect of vitamin addition had little effect on oxygen consumption. The authors conclude that whatever decrease in oxygen consumption is present in the cerebrum, it is not due to vitamin B1 decrease, as it is not reliably reversed by the addition of vitamin B1.

Adding vitamin B1 to a regional brain preparation seemed to rule out any stimulation in vivo from some other agent or hormone in vitamin-deficient pigeons. Similarly, these data correlate well with the lactate data gathered earlier. Collectively, the data on lactate and oxygen consumption indicate a local (regional) biochemical lesion, which correlates with an abnormal condition in the CNS. The authors conclude that vitamin B1 is directly involved with oxidative metabolism in lower brain parts in symptomatic pigeons.

In a later paper (Rydin, H., 1935), pigeons that were not in opisthotonus but rather were ataxic due to vitamin B1 deficiency were studied. The goal was to see if there was a change in oxygen uptake in regional brain parts when other symptoms were present.

As before, pigeons were made symptomatic by feeding them polished rice. Birds were on this diet for at least 30 days and were selected for the study based on the absence of opisthotonus and presence of leg weakness and ataxia. Animals were killed by decapitation and brains quickly removed and minced with a spatula. Oxygen consumption was then measured in the cerebrum and in the lower parts of the symptomatic pigeons' brains.

Results showed that pigeons exhibiting leg weakness had a greater increase in oxygen uptake due to added vitamin B1 in the cerebrum. There was a negligible effect in the optic lobes and the rest of the brain. In previous work the optic lobes and the rest of the brain were affected during the opisthotonus phase of vitamin B1 deficiency.

In this study, as seen earlier, the decrease in oxygen uptake and symptoms correlated with vitamin B1 deficiency, not inanition. This study shows this even in the absence of opisthotonus, but with leg weakness, a decreased oxygen uptake was reversed by vitamin addition in vitro, again arguing against the stimulation of some other hormone or chemical release. Finally this study demonstrates that a different brain region – the cerebellum – appears to respond to leg weakness. This shows that the heterogeneity of the brain accounts for different responses, which correlate with different neurological symptoms. This study points out again that the leg weakness was central in origin and not manifest due to altered muscle biochemistry or peripheral nerve involvement.

In a subsequent paper the oxygen/pyruvate ratio and the pyruvate RQ in normal and vitamin-deficient pigeon brain (McGowan, 1937) were determined. The methodology was as previously described from the Peters' laboratory.

Results showed that pyruvate was incompletely oxidized, and it was likely that some intermediate was accumulating. Also it seemed that some pyruvate was completely oxidized, while some was metabolized along some other route. Data from another worker showed that some pyruvate was metabolized to lactic acid and CO_2. Assuming that two-thirds of pyruvate is completely oxidized, and one-third goes into the Krebs cycle, then the theoretical oxygen/pyruvate ratio is around 450 and the RQ is 1.29, and these are about the actual values found.

The experiments described above were carried out in pigeons suffering from vitamin B1 deficiency due to eating polished rice. Results showed that the oxygen/pyruvate ratio was significantly lower than that of normal pigeons. The author suggests that there may be two different routes for pyruvate, and absence of vitamin B1 may influence which route is utilized in vitamin B1 deficient birds.

In a neuropathological study, pigeons were made thiamine deficient by dietary means, then brains and peripheral nerves were examined in animals exhibiting opisthotonus, leg weakness, and in controls without symptoms (Swank, R., 1940). Experimental groups consisted of the following: (1) starved and thiamine deficient, (2) acutely thiamine deficient, (3) chronically thiamine deficient, and (4) starved on normal diet. Acute thiamine deficiency was produced by tube feeding a deficient diet until vomiting occurred. Then thiamine was administered along with the diet for 3–6 weeks. After that, symptoms (opisthotonus) occurred in 7–12 days if the thiamine supplementation was stopped. In the chronic group, after vomiting occurred, a small dose (7–15 μg) of thiamine was administered. In these birds, ataxia and leg weakness occurred between the 16th and the 30th day.

Results in the acutely thiamine-deficient group showed occasional degeneration in the sciatic nerve only. Lesions were also seen in the ventral funiculus of the cervical spinal cord. The lesions were those of degenerating myelin sheaths. It was thought that there were more affected nerves, but that the chlorate-osmic acid staining method would not show them.

In the chronically thiamine-deficient group, pigeons developed leg weakness and ataxia, not opisthotonus. In addition most of these pigeons showed evidence of heart failure. Results showed that the greater and longer lasting symptoms correlated with more pronounced evidence of demyelination in peripheral nerves. In mildly symptomatic birds, the pathology was distal, and as the intensity of symptoms grew worse, the lesion moved centrally. The process of degeneration was Wallerian in appearance. Centrally, lesions start in the lateral funiculus and follow the spinocerebellar tract and can be traced into the cerebellum. This "ascent" was similar to that seen peripherally where lesions first appeared distally and moved centrally in increasingly severely symptomatic birds.

In birds "reversed" with thiamine treatment, the total number of myelinated nerve fibers in the sciatic and brachial nerves increased as the birds recovered. Complete recovery took 6–9 weeks in birds in which leg weakness had lasted for 7–10 days before instituting treatment. This is in contrast to the rapid recovery (hours) in rats reversed with thiamine therapy. Microscopically, dorsal ganglia continued to show chromatolysis until symptoms cleared. It seemed that complete regeneration of nerve fibers had to occur before symptoms disappeared.

The author suggests that the lesion associated with opisthotonus is biochemical in nature, and that explains the rapid reversal of symptoms when the symptom is opisthotonus. By contrast, the leg weakness appears to be a lower motor neuron type of paralysis, and nerve fibers in the sciatic nerve degenerate. Adding thiamine to the diet reverses the lesion and symptoms, but the symptom reversal is dependent upon reversal of the nerve lesion. Hence, this reversal time is lengthy. The large myelinated peripheral nerve fibers degenerated first in thiamine-deficient birds. In the CNS, the long ascending spinocerebellar fibers were affected early in the process. When an axis cylinder has been damaged for longer than 4 days, changes in the cell bodies consisting of chromatolysis occur. The chromatolysis remains after initiation of the thiamine reversal period and finally disappears after the myelin cylinder shows complete repair. In the CNS, damaged neurons did not regenerate. It is suggested that in early thiamine deficiency, CNS neurons are at first damaged only chemically. If thiamine is administered during this early period cell recovery is about 100%. Later, structural changes occur in the cell body, which disappear only if the axis cylinder makes a full recovery.

Opisthotonus is a manifestation of decerebration. Leg weakness, a more chronic disorder, is due to degeneration of nerves such as the sciatic nerve. Reversal is possible given thiamine treatment and time.

The blood lactate/pyruvate relation in thiamine-deficient pigeons was investigated (Stoltz, E. and Bessey, O., 1942). Pigeons were fed the usual polished rice diet and blood was collected from a wing vein. Birds were on a thiamine-deficient diet for 9 days. At this point, blood sampling revealed increases in pyruvate, at a time when the pigeons were still asymptomatic. The birds were then allowed 5 more days of thiamine-deficient diet, at which time some pigeons were exhibiting mild opisthotonus. This was the time at which the pyruvate values were greatest. Treatment of symptomatic pigeons with thiamine reversed these elevated pyruvate levels. This study demonstrated the usefulness of blood pyruvate/lactate measurements as a way to evaluate thiamine deficiency even before the onset of symptoms.

In an early study in rats (Harper, H., 1942), the rate of absorption of glucose from the intestine was examined. Rats were placed on a thiamine-deficient diet and maintained in a way that allowed a "subacute" thiamine-deficient state to exist. At the time of killing, glucose was administered, and after 1–2 hours, the entire digestive tract was removed and analyzed for glucose. Results showed that there was a significant decrease (17%) in glucose absorption. This decrease in glucose absorption was not surprising in light of the decreased function in the GI tract. In another series of experiments, the author found decreased glycogen in the liver of thiamine-deficient rats 3–6 hours after feeding glucose. This finding could be attributed to lower absorption of glucose.

The effect of oxygen and thiamine pyrophosphate (cocarboxylase) on the formation of citrate and alpha-ketoglutarate has been studied (Coxon, R. and Peters, R., 1950). In this study, pigeons were used and treated by administering polished rice as the diet until opisthotonus developed. Animals were killed and brain homogenates were prepared. Both citrate and alpha-ketoglutarate were measured. Results showed that the addition of cocarboxylase and oxygen to pigeon brain extracts promoted the generation of citrate and alpha-ketoglutarate. In pigeon brain preparation from thiamine-deficient animals, pyruvate accumulates. Previously it had been suggested that some seven-carbon compound was the first compound formed in the decarboxylation of pyruvate. This seems not to be the case given that it is pyruvate which accumulates when citrate and alpha-ketoglutarate do not form in thiamine-deficient pigeon brain. Thus it is shown that citrate and alpha-ketoglutarate are formed in the presence of oxygen and cocarboxylase. Further, it is concluded that a two-carbon compound is formed when pyruvate is decarboxylated.

Based on earlier work showing that thiamine deficiency interferes with the degradation of pyruvate, the author of this paper (Liang, C., 1962) examined pyruvic acid changes in thiamine deficiency.

Albino rats were made thiamine deficient by feeding a diet of polished rice powder. Pyruvic acid and total keto acids were measured. Results showed that rats on a thiamine-deficient diet died after 10 weeks, and they demonstrated weight loss and ataxia.

Pyruvate and keto acids in brain, heart, liver, and muscle showed a steady rise after administration of the deficient diet. Glyoxylic acid was detected in all tissues tested, including brain.

A slight decrease was seen in the rise in pyruvic acid and keto acids between days 15 and 22, and the reason is unclear. The reason for the elevation in brain of glyoxylic acid also remains unclear, but may be related to the deamination of glycine and the amination of alpha-ketoglutarate to glutamate and glutamine. Glyoxylic acid is toxic. These data are in keeping with previous reports showing impairment in enzyme activity of pyruvate decarboxylase and of alpha-ketoglutarate decarboxylase in the brains of thiamine-deficient rats.

In a paper by Lofland et al. (1963), two different strains of pigeons were studied as to the effect of thiamine deficiency on thiamine-dependent enzymes. Results showed that the "snow racer" pigeon was more susceptible to thiamine deficiency than the "white carneau" breed. This was viewed as an indication of a possible genetic control of thiamine-dependent enzymes. The enzyme alpha-ketoglutarate decarboxylase seemed to show the most decreased activity among the three enzymes studied. Results also showed that transketolase decreased in blood, but remained about normal in brain. The activity of pyruvate oxidase was decreased, which seemed to correlate with the elevation in pyruvic acid. Overall these data are in keeping with previous data showing an alteration in the thiamine-requiring enzymes in carbohydrate metabolism (for further review, see Itokawa, 1995).

Thiamine Deficiency in Mammals

The use of animal models of thiamine deficiency has proved quite beneficial in the elucidation of underlying mechanisms of pathogenesis. Brain lesions, time course, symptoms, and reversibility of symptoms all correlate very well with features seen in humans. The issue of mode of action of thiamine antagonists has been investigated in thiamine-deficient rats (Gubler, 1961). Although several antagonists have been discovered such as oxythiamine, pyrithiamine, amprolium, and deaminothiamine most studies have used either oxythimine or pyrithiamine. The Gubler study examined the effects of a pure dietary thiamine-deficient diet, a thiamine-deficient diet plus oxythiamine, and a thiamine-deficient diet plus pyrithiamine on the oxidation of rat brain pyruvate and alpha-ketoglutarate.

Thiamine-deficient rats on this regime showed the usual growth curve (see Fig. 1, for example). After sacrifice, brain homogenates were prepared and the oxidation of pyruvate and alpha-ketoglutarate measured in brain homogenates. Results from these biochemical analyses showed that the oxidation of neither pyruvate nor

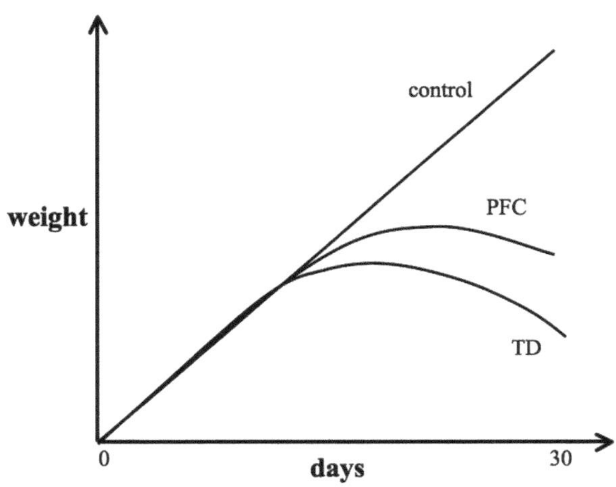

Fig. 1

D.W. McCandless, *Thiamine Deficiency and Associated Clinical Disorders*, Contemporary Clinical Neuroscience, DOI 10.1007/978-1-60761-311-4_3,

alpha-ketoglutarate was decreased by oxythiamine, but both were decreased with thiamine deprivation and pyrithiamine treatment. These decreases could be reversed by the addition of thiamine diphosphate in vitro.

Heart enzymes were also measured, and pyruvate oxidation was decreased in all three types of thiamine deficiency, but only oxythiamine plus thiamine-deficient diet had a lowering effect on alpha-ketoglutarate oxidation. These data indicated that at least two enzymes are involved in the oxidation of pyruvate and alpha-ketoglutarate.

In a separate study (Brin, 1962), the effects of thiamine deficiency plus oxythiamine was examined on rat brain transketolase. This study showed that in all tissues examined – kidney, intestine, spleen, heart, liver, etc. – transketolase levels were significantly lower. The one exception was brain. The decreases were evident in tissue before the leveling off of the growth curve. The level of decrease in transketolase with oxythiamine administration was noteworthy in that, at that time period (13 days), neurological symptoms were absent.

In another study (Dreyfus, 1965), transketolase was measured in brain regions of thiamine-deficient rats maintained on diet for 30+ days. Brain regions assayed included myelinated areas such as spinal cord and pons and less myelinated areas such as cerebral cortex and thalamus. High transketolase levels were found in spinal cord and lowest levels in the cerebral cortex. The greatest effect of thiamine deficiency was noted in the lateral pontine tegmentum, the site of the most striking neuropathological changes. These data suggested that transketolase activity might be involved in oligodendroglial metabolism.

A problem of older studies using Warburg techniques to measure pyruvate dehydrogenase activity is a lack of sensitivity. To overcome this, a study was performed comparing four different measures of pyruvate dehydrogenase activity: (1) Warburg manometry, (2) the Clark electrode method, (3)14CO_2 evolution, and (4) ferricyanide reduction (Bennet et al., 1966). Tissue analyzed was liver and brain from pyrithiamine-induced thiamine-deficient rats. Animals were maintained on this diet until symptomatic and thenkilled. Results showed that all four procedures for estimating pyruvate dehydrogenase activity gave similar results. The authors found no effect of thiamine deficiency on liver enzyme activity. Brain pyruvate dehydrogenase activity was reduced about 70% compared to control rats. Cerebral 2-oxoglutarate oxidation was reduced by 30%, whereas succinate and malate oxidation was not changed (Fig. 2).

In a study comparing the oxidation of pyruvate and the oxidation of ribose in intact rats (Brin, 1967), 14C-pyruvic acid and 14C-ribose were injected into thiamine-deficient rats, and 14CO_2 evolution measured. Results showed that 14C-pyruvate was oxidized to 14CO_2 in deficient rats at the same rate as controls. 14C-ribose, however, showed a reduced oxidation at only 1 week of deficiency and continued to decrease throughout the time course. The defect showed improvement within 24 hours following thiamine treatment. Transketolase activity was measured in six different tissues in thiamine-deficient rats and found decreased in kidney, muscle, and liver, but not brain.

The use of the thiamine antagonists oxythiamine and pyrithiamine became increasingly widespread, and the mode of action of these antagonists was examined

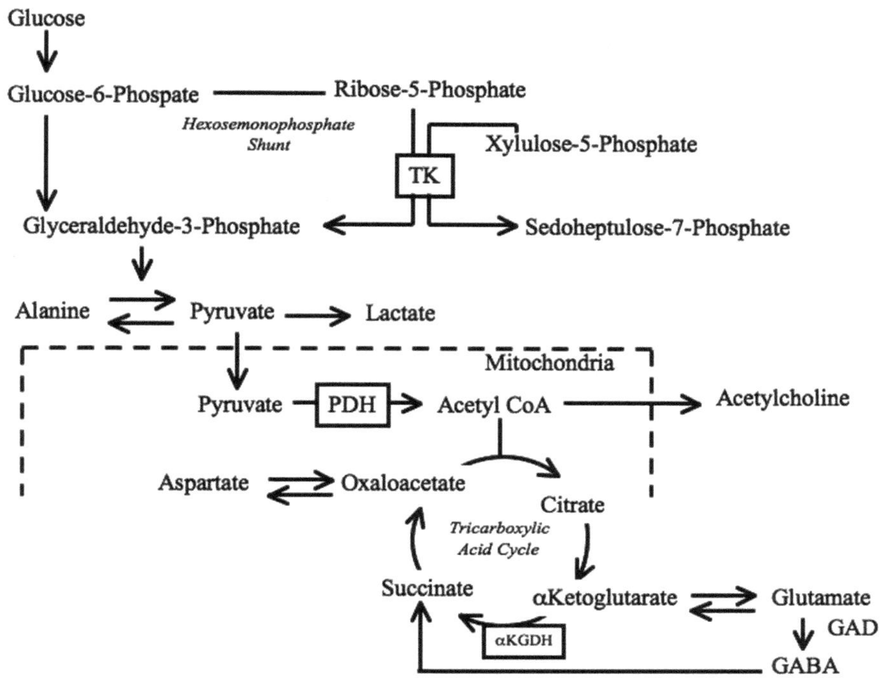

Fig. 2 Schematic representation of glycolysis and the TCA cycle. Modified from Desjardins, P. and Butterworth, R. (2005), and McCandless, D. and Schenker, S. (1968)

(Steyn-Parve, 1967). When antimetabolites are administered, many symptoms noted in pure dietary thiamine deficiency are present. These effects can be reversed by increasing thiamine doses. The effects on mice and rats of the two antagonists are different. Oxythiamine seems to alter heart function, whereas pyrithiamine inhibits thiamine phosphate synthesis and has its major effect on the nervous system. Oxythiamine acts as a competitor of thiamine pyrophosphate. Oxythiamine produces bradycardia, elevates blood pyruvate significantly, and cannot enter brain tissue. Pyrithiamine is capable of entering brain tissue, thereby producing CNS symptoms. These studies tend to dissociate the heart and CNS symptoms in thiamine deficiency by showing that the two clinical characteristics can exist independent of each other.

Additional studies (for review, see Cooper and Pincus, 1967; and Karuppagounder, S. and Gibson, G., 2009) were directed toward determining if thiamine (or thiamine triphosphate) might play a role in nerve conduction. Initial studies had shown that pyrithiamine applied to nerve fibers in vitro blocked the action potential, whereas oxythiamine did not. Another key observation was that in vitro frog spinal cord released thiamine when stimulated. Other studies showed that vagus nerve preparations to which pyrithiamine was administered showed an increase in amplitude of the compound action potential and reversal of post-tetanic hyperpolarization. These changes could be reversed by thiamine administration, and oxythiamine did not have an effect on vagus nerve electrical activity. Results on experiments to analyze the effects of pyrithiamine on vagus nerve preparations were equivocal.

To determine if thiamine might be involved in nerve conduction, a further look at the mode of action of pyrithiamine was undertaken. In one set of experiments, vagus nerves were incubated for 10 minutes both with and without pyrithiamine. Nerves incubated with pyrithiamine had about a 40% decrease in thiamine as compared to those not incubated with pyrithiamine.

In a second series of experiments, rats were made deficient by dietary means. When symptomatic, labeled thiamine was injected over 3 days. Upon killing, spinal cords were removed, divided longitudinally, and placed on electrodes in a Locke's bath. The spinal cords were stimulated in the bath both with and without pyrithiamine. Results showed that both pyrithiamine and electrical stimulation released labeled thiamine from the spinal cord preparations. This correlates with earlier pyrithiamine studies showing an increased urinary output of thiamine during pyrithiamine administration. These data, overall, suggested a possible nerve conduction role for thiamine, independent of its role as co-enzyme in key enzymatic reactions. Direct evidence of participation of thiamine or thiamine triphosphate in nerve ion transport is lacking.

The possibility that there was a relationship between thiamine deficiency and magnesium deficiency was suspected on clinical grounds. Therefore, studies were undertaken to examine possible correlations (Zieve et al., 1968). Results showed that in magnesium-deficient rats, blood and liver thiamine levels were slightly reduced, and in thiamine-deficient rats, there was a transient decrease in magnesium in liver. The activity of thiamine-requiring enzymes was significantly decreased in the liver of magnesium-deficient rats. The recovery of enzyme activity was incomplete upon thiamine administration. Interestingly, magnesium in vitro did not reverse decreased enzyme activity.

The explanation that seemed most logical to the authors was that magnesium deficiency interfered with activation of magnesium-dependent enzymes. The effect on transketolase, rate limiting reaction in the hexose monophosphate shunt, seemed most sensitive. The authors suggest a defective incorporation of magnesium into the enzyme complex. These data are somewhat in keeping with earlier studies (Brin, 1967) suggesting the transketolase alteration in thiamine deficiency is more critical than that of the decarboxylation of pyruvate or oxyglutarate.

The authors speculate that a state of magnesium deficiency, appearing clinically similar to thiamine deficiency, could explain why thiamine alone does not ameliorate symptoms. The authors, however, found it strange that Mg replacement therapy did not result in a reversal of deficiency symptoms. This was interpreted as further evidence of an effect on transketolase and that this effect carries over to the recovery phase. Both animals and humans could be compromised by magnesium deficiency because it would result in a reduced ability to properly utilize thiamine.

The effect of thiamine deficiency on rat brain acetylcholine has been examined (Cheney et al., 1969). Thiamine-deficient rats treated with the antagonist pyrithiamine developed neurological symptoms and seizures. There was a lack of correlation between these symptoms and either total acetylcholine activity or between symptoms and the rate of acetylcholine formation in the deficient rat brains. On the other hand, prevention of acetylcholine destruction by injection of eserine prior to convulsions seemed to prolong death. This suggests that perhaps acetylcholine does play some (minor) role in the evolution of neurological signs.

Most of the above studies were carried out in whole brain, but neuropathological studies in thiamine deficiency indicate a highly focal lesion in the brainstem and cerebellum, as is seen in Wernicke's disease. When analyzing whole brain, inclusion in the sample of non-affected tissue could easily dilute the sample, thereby masking adjacent affected brain areas.

In a study aimed at looking at biochemical parameters in brain regions (McCandless and Schenker, 1968), thiamine deficiency was produced by dietary methods only. This regime produces neurological symptoms in 4.5–5 weeks. Since loss of appetite and decreased food ingestion is a feature of dietary thiamine deficiency, pair-fed controls were utilized in order to have weight loss controls. Animals were killed in a dry ice acetone mixture (–120°C) in order to preserve labile brain metabolites.

By 4.5 weeks, thiamine-deficient rats began to show neurological symptoms. These consisted of ataxia, impaired righting response, opisthotonic posturing, and drowsiness. Seizures were not a feature of this dietary regime. When the deficient rats were injected with 10 μg of thiamine, symptom reversal occurred within 16–36 hours. The ability of the rats to be "reversed" allows for the brain biochemical changes to be re-examined following symptom reversal.

Cerebral thiamine levels dropped to about 15% of control values when symptoms were most severe. Thiamine dropped to 20% of control before any symptoms occurred, indicating a significant brain reserve of the vitamin. There was no difference between ad lib fed controls, and pair-fed controls.

The thiamine dependent enzyme pyruvate decarboxylase was decreased by 36% and 29%, respectively, in brainstem and cerebellum of symptomatic rats. Reversal of symptoms with B1also reduced the enzyme depletion. The thiamine-requiring enzyme transketolase was also measured, and the most significant decrease was found in thiamine-deficient cerebral cortex, an area not affected in thiamine deficiency. The enzyme decrease was least in the brainstem of deficient rats.

Lactate levels were elevated in the cerebellum and brainstem of thiamine-deficient rats. They were unchanged in the cerebral cortex. Brainstem pyruvate was also elevated in symptomatic animals. Reversal of symptoms with thiamine administration reduced the decreases in these metabolites. Analysis of ATP levels in cortex, cerebellum, and brainstem showed no alteration in ATP in moderately or severely thiamine-deficient rats.

This study defined the critical range of thiamine depletion necessary to result in neurological symptoms, and the amount of increase related to the reversal of symptoms with thiamine treatment. Transketolase was decreased to a greater extent than pyruvate decarboxylase, duplicating previous studies. Regional results showed a greater decrease in brainstem/cerebellum than in cerebral cortex. These findings are consistent with the neuropathologic observations of lesions largely localized to the cerebellum and brainstem.

The rapid reversal (16–36 hours) of symptoms with thiamine administration permits an opportunity to re-measure enzyme levels to determine if one or the other also reverts toward normal. In spite of the greater effect of thiamine deficiency on transketolase levels, it was the pyruvate decarboxylase enzyme that had a more dramatic increase toward normal. This would suggest that reversal of the pyruvate decarboxylase enzyme activity was more closely tied to the reversal of symptoms.

Since ATP production is linked to pyruvate decarboxylase and TCA cycle function, the finding of normal ATP levels suggests that some other mechanism linked to the enzyme level should be examined. This does not mean that some smaller region might show changes in high-energy phosphates or that there could be a decrease in production of ATP coupled with a decrease in utilization leading to unchanged net levels of high-energy phosphates. Remember, the lesions of thiamine deficiency are highly localized.

Utilization of the Lowry "closed system" (Lowry et al., 1964) assesses turnover of adenine nucleotides and has been used to examine flux in thiamine-deficient rat brain (McCandless and Cassidy, 1976). In this study, after decapitation, the brains are maintained at 37°C for 30 seconds and then frozen in liquid N_2. This effectively stops high-energy phosphate synthesis, and only utilization occurs (Lowry et al., 1964). Results from this study showed that metabolism of the adenine nucleotides ATP, ADP, and AMP were not altered in severely thiamine-deficient rats. Pyruvate decarboxylase activity was also measured and found to be 63% decreased in the brainstem of thiamine-deficient rats as compared to pair-fed controls. This indicates that there is a significant reserve capacity for pyruvate decarboxylase.

The above studies, although performed regionally, still included many cell types, which may not be involved in severe thiamine deficiency. To circumvent this problem, cerebral metabolism in thiamine-deficient rats has been examined in small (100–500 ng) samples from the lateral vestibular nucleus (McCandless and Schwartzenburg, 1981).

The results from this study showed that phosphocreatine was elevated in the lateral vestibular nucleus in severely thiamine-deficient rats as compared to pair-fed controls. Reversal of symptoms resulted in a return of phosphocreatine to control levels. An increase in phosphocreatine could be viewed as an indication of a decrease of energy utilization rate in very discrete samples. ATP was unaffected in this study. Glucose and glycogen levels increased in the recovery stage, a finding noted in the reversal of other models of metabolic encephalopathy (McCandless et al., 1979).

It has been shown previously that transketolase is depleted in thiamine-deficient brain to a greater extent than pyruvate decarboxylase. This observation has led many to speculate that transketolase and the hexose monophosphate shunt play a vital role in thiamine deficiency. This hypothesis has been advanced despite the fact that upon reversal of symptoms with thiamine administration, transketolase levels returned toward normal only slightly, whereas pyruvate decarboxylase returns much more dramatically. Interestingly, thiamine triphosphate (with a supposed neurotransmission role) returns not at all (Pincus and Grove, 1970). Since enzymes are not a true indicator of flux through a metabolic pathway, the effect of thiamine deficiency directly on the hexose monophosphate shunt (pentose phosphate pathway) was examined.

In these studies (McCandless et al., 1976a), rats were rendered thiamine deficient to the point of moderate to severe neurological signs. Flux through the hexose monophosphate shunt was estimated based on the differentially labeled glucose method. The C1/C6 ratio derived from the specific yields of $14CO_2$ from 1 to 14C and from 6 to 14C glucose shows that there was no decrease in flux through the pathway. In fact there was a 20% increase in flux through the shunt. This probably represents a compensatory mechanism that is not understood. The effect of thiamine deficiency on transketolase in this study was a 65% decrease in activity. There are some limitations to this method (Katz–), and whole brainstem was used, so a more anatomically discrete study might yield different results. This study does not, however, argue in favor of a significant role for the hexose monophosphate shunt in the pathogenesis of the symptoms of thiamine deficiency.

Pyrithiamine-induced thiamine deficiency can be produced in mice in 10–12 days, and this model was the basis for an electron microscopy study (Watanabe et al., 1981). Groups of mice included the following: (1) those made thiamine deficient using a thiamine-deficient diet plus pyrithiamine; (2) mice treated as in group 1, then reversed by injection of thiamine; and (3) mice made thiamine deficient as above, then reversed, and then maintained on a thiamine-containing diet for 2.5 weeks. Lesions in the brains of thiamine-deficient symptomatic mice were localized in the thalamus, pontine tegmentum, and mammillary bodies. These lesions were characterized by edematous necrosis localized to neuron dendrites, myelin sheaths, and astrocytes. Also present was hemmorhagic necrosis. Following reversal of symptoms (48–96 hours), edema was still present, and fat-laden macrophages were noted. After 2.5 weeks of "reversal," some edema was still present, as well as Wallerian type of degeneration. It was noted that the lesions resembled those of non-hemmorhagic Wernicke's disease. Vascular permeability to horse radish peroxidase was seen in severely affected mice, but it reversed when the symptoms were reversed with thiamine treatment. The findings suggest that both nerve and vascular lesions are present, but that the nervous lesions are more significant than the vascular changes.

Another study examined neuropathological findings in a series of dogs made thiamine deficient by dietary means (Read and Harrington, 1986). Lesions occurred in two varieties: (1) in this group, lesions were confined to the inferior colliculus, and (2) a group consisting of much more widespread lesions in the cerebral cortex, inferior colliculus, cerebellar nodulus, and medial vestibular nucleus. In both groups, gray matter was the site of lesions, and the lesions were bilaterally symmetrical. Microscopically, the lesions were varied, but included spongiosis, vacuolation of neuropil and myelin sheaths, neuronyl necrosis, hypertrophy and hyperplasia of endothelial cells, glial necrosis, and accumulation of lipid containing phagocytes.

The presence of two different neuropathological patterns was seen as a different chronic pattern and different symptoms in the affected dogs. The range of time for the development of overt thiamine deficiency was 32–134 days.

Another study (Aikawa et al., 1984) examined the neuropathological and biochemical effects of an acute onset of pyrithiamine-induced thiamine deficiency. Wistar rats were used in this study, and results showed the usual weight loss, and by days 11–12, a rapid onset of symptoms including ataxia, opisthotonus, tonic seizures, etc. Grossly, the brains appeared normal except for occasional petechial hemorrhages. Results from light microscopic examination showed spongy changes in the thalamus, mammillary bodies, and lateral vestibular nucleus in the pons. These changes were bilaterally symmetric. Except for severe cases, lesions were limited to astrocytes and oligodendrocytes. In severe cases neurons and myelin sheaths were involved. Petechial hemorrhages were infrequently seen. Electron microscopic findings were consistent with those seen in light microscopy.

Biochemical analysis included measuring thiamine and high-energy phosphates. Results showed that cerebral thiamine dropped to less than 20% of control levels, a finding previously described (McCandless and Schenker, 1968). High-energy phosphates were decreased in symptomatic animals; ATP was down 10% and phosphocreatine decreased by 30%. These data are somewhat at odds with other investigators' results, but the rats in this study were experiencing tonic seizures. Seizures are known to have a deleterious effect on cerebral energy metabolism (McCandless et al., 1979). Most previous studies on thiamine-deficient animals stress that seizures were not a clinical feature of murine thiamine deficiency. Therefore, care must be taken in the interpretation of the high-energy phosphate data in the paper by Aikawa et al. (1984).

Another study looking at development examined the effects of pyrithiamine-induced thiamine deficiency on rat brain myelination (McCandless et al., 1976b). Newborn rats were injected with a low or high dose of pyrithiamine, then killed at various times and brains analyzed for transketolase and pyruvate decarboxylase, and brains were stained for myelination, thought to be dependent on the hexose monophosphate shunt.

Results were that pyruvate decarboxylase, as indicated by pyruvate levels, were depleted by about 30%. Similarly, transketolase levels were depleted by as much as 36% in high-dose rats. Staining of brains by the Weil technique showed no alteration in amount or timing of myelin deposition. Myelin components such as phospholipids, cholesterol, and cerebrosides were not depleted in experimental animals. The newborn rats were symptomless and had normal weight gain. These data indicate that in spite of an enzymatic defect that is lethal in adult rats, the newborn rats develop normally, and their myelination is normal. This is interpreted as suggesting alternative pathways supporting myelination were available to newborn rats or that some heretofore undescribed protective device is in place permitting myelination.

Cerebral blood flow in thiamine-deficient rats has been measured regionally (Hakim, 1986). 14C-iodontipyrine was used to image and measure regional cerebral blood flow in rats rendered thiamine deficient using a thiamine-deficient diet plus pyrithiamine injections. Animals became anoxic and started loosing weight after day 10. By day 18, animals suddenly started showing neurological symptoms such as ataxia, opisthotonus, and loss of righting response. PO_2 levels were decreased, and symptomatic rats were hyperglycemic. 14C-iodoantipyrine autoradiographs showed evidence of early hyperfusion, especially in areas known to be affected in thiamine deficiency, such as the thalamus and inferior colliculus. There was a delayed hypoperfusion seen later in the pathologic process. The author attributes these local blood flow changes to metabolic changes in local cerebral glucose utilization described earlier (Hakim and Pappius, 1981).

Acetylcholine has been measured in the brains of thiamine-deficient animals with divergent results. The rationale for looking at acetylcholine is that its production might be decreased either by interference with acetyl-CoA production via pyruvate decarboxylation activity depletion or by decreased production due to lowered ATP levels.

In the study described (Vorhees et al., 1977), normal controls and pair-fed controls were compared to thiamine-deficient rats in terms of acetylcholine levels and also turnover of acetylcholine. Microwave irradiation was used for killing; this technique minimizes any chance for decay of labile metabolites before assay. Animals were made thiamine deficient using dietary means, and pair-fed controls were fed normal diet in the amount the thiamine-deficient rats consumed the previous day.

Acetylcholine was measured in several brain regions, including cerebral cortex, diencephalon, midbrain, cerebellum, and medulla-pons. Results for net acetylcholine levels showed no significant differences in levels in any region in deficient rats as compared to either pair-fed controls or rats fed a normal ad lib diet. Utilization of acetylcholine following inhibition of synthesis with hemicholinium-3 showed a small but significant regional decrease in utilization. This decrease was only statistically significant in the midbrain. The data, taken together, point to impaired synthesis of acetylcholine. If one supposes decreased synthesis of acetylcholine, the mechanism may not be through acetyl-CoA since it has been shown to not be altered in thiamine deficiency (Reynolds and Blass, 1975).

Since transketolase, a thiamine-requiring enzyme, is rate limiting for the hexose monophosphate shunt, lipids and myelination have been investigated in young developing rats (Geel and Dreyfus, 1975; McCandless and Cassidy, 1976). In the first study, the lipid composition was studied in brains of 25-day-old rats made thiamine deficient from gestational day 14. These experimental animals displayed weight loss and neurological symptoms such as tremor, opisthotonus, and ataxia. The undernutrition pair-fed control rats had a reduction in brain lipids. Thiamine-deficient rats had no further change in regional brain lipids as compared to pair-fed controls. The authors interpret these findings to cast doubt on the hypothesis that depleted transketolase limits the hexose monophosphate shunt's ability to generate reducing power (NADPH) for lipid synthesis. In general agreement with these results is a study looking at myelination in thiamine-deficient newborn rats (McCandless et al., 1976b). This study showed normal myelination throughout the critical period for myelin deposition, in spite of thiamine deficiency.

In another interesting study (Mesulam et al., 1977), rhesus monkeys were gradually weaned onto a thiamine-deficient diet. When the deficient group became severely affected, they were reversed with thiamine and then made deficient again. Clinical manifestations were carefully monitored, and blood transketolase monitoring confirmed thiamine deficiency.

Clinical signs and symptoms included apathy, anorexia, nystagmus, ataxia, and heart failure. Onset of these symptoms ranged from 40 to 100 days. These symptoms were reversed with thiamine administration for several days and then the animals were again made thiamine deficient. Several episodes of this regime were performed, and clinical signs tended to worsen with each episode. This progression may have been due to slowly developing brain damage.

This study is reminiscent of the history of chronic alcoholics, and the signs and symptoms displayed by these thiamine-deficient rhesus monkeys were strikingly similar to those seen in the Wernicke-Korsakoff syndrome. Transketolase levels in blood confirmed chemically that the monkeys were indeed thiamine deficient. The authors make the point that the treatment of chronic alcoholics in hospitals with large thiamine doses may increase variations in thiamine levels. The present study in monkeys shows that reversal of symptoms with large thiamine doses may make subsequent deficiency status worse clinically. There is a possibility that similar "up and down" thiamine levels in humans may have a deleterious effect.

The effect of thiamine deficiency on the neurotransmitter serotonin has been examined (Plaitakis et al., 1978). The rationale was that evidence existed for a possible role of thiamine in nerve conduction, as described above. In this study, both pyrithiamine and a low thiamine diet were used to produce thiamine deficiency. Results showed that other neurotransmitters studied such as GABA, glutamic acid, norepinepherine, and choline were not affected by thiamine deficiency. Serotonin, however, was affected in that synaptosomes from the cerebellum of thiamine-deficient rats showed the Vmax for serotonin uptake was only 50% that of controls.

The decreased uptake of serotonin was limited to the cerebellum; other regions were not affected. Thiamine added in vitro did not restore decreased serotonin uptake; however, when thiamine was added in vivo, reversal of symptoms occurred, as well as reversal of decreased serotonin uptake. Thiamine deficiency in rats produces seizures in some studies, and hypothermia in all studies. These two symptoms alone are associated with decreased serotonin uptake. Serotonon uptake is energy dependent and there are some mild effects of thiamine deficiency on energy metabolism. Taken together, it seems possible that the changes in serotonin in thiamine-deficient rat brain are secondary to some other key changes.

A close relationship exists between the excitatory amino acids glutamate and aspartate, and the thiamine dependent enzyme alpha-ketogluterate dehydrogenase. This prompted investigators to examine the effect of pyrithiamine-induced thiamine deficiency on regional cerebral glutamate and aspartate metabolism (Butterworth and Heroux, 1989). Measurement of these two amino acids in symptomatic pyrithiamine-induced and symptomatic thiamine-deficient rats showed decreases in glutamate in the thalamus and pons and decreases in aspartate in the thalamus, pons, cerebellum, and cerebral cortex. The enzyme alpha-ketoglutarate dehydrogenase was decreased regionally, whereas activities of pyruvate dehydrogenase were unchanged.

Upon reversal of symptoms with thiamine administration, concentrations of glutamate and aspartate, and alpha-ketoglutarate returned toward normal. Alanine (also measured) levels increased in the areas studied during the symptomatic stage and returned to normal with thiamine treatment. Alanine is often associated with pyruvate decarboxylase activity, but the enzyme was unchanged in this study.

Pyrithiamine is used to produce thiamine deficiency, coupled with thiamine-deficient diet. The advantage of using the antagonist is that symptoms are usually produced in 10–12 days, as compared to 30+ days required with thiamine-deficient diet alone. Another advantage is that the symptoms are compressed into a day or so, whereas onset of symptoms generated by dietary means alone might take 3–4 days. This makes the pyrithiamine-induced thiamine deficiency model more efficient. It is, however, a different model as shown clearly in the above study. For example, pyruvate decarboxylase activity is unaltered in the above pyrithiamine model, and it is always changed in the dietary method. When making correlations to human syndromes such as the Wernicke-Korsakoff disease, one should be careful to use a murine model (dietary) most closely resembling the human counterpart.

In another study (Fournier and Butterworth, 1990), thiamine-dependent enzymes were measured in thiamine-deficient pregnant rats and their offspring. Maternal rats had decreased transketolase levels in the brain regions cerebral cortex, cerebellum, and brain stem in thiamine deficiency lasting 22 days. Pyruvate dehydrogenase and alpha-ketoglutarate dehydrogenase activities were unaltered in maternal rats in the three regions examined. Neurological symptoms were not mentioned in this paper, but symptoms do not occur in dietary-induced thiamine deficiency until around day 28. Red blood cell transketolase activation (TPP effect) was used to assess thiamine deficiency, and it was decreased.

The offspring at 13 days of age showed decreases in pyruvate dehydrogenase and alpha ketoglutarate dehydrogenase in the cerebral cortex, and transketolase was decreased in all three regions: cerebral cortex, cerebellum, and brainstem. These changes could certainly have deleterious effects on brain development during critical periods for metabolic activity. The peculiar findings as regards the distribution of effects on pyruvate dehydrogenase and alpha-ketoglutarate dehydrogenase remain unexplained, but could be a non-descript early deficiency change. Transketolase activity changes might affect lipid/myelination metabolism; however, earlier studies show no effect on myelin formation, and flux through the hexose monophosphate shunt is not affected by a moderate drop in transketolase activity.

A method for producing murine pyruvate dehydrogenase deficiency has been developed in pregnant mice. The mutation is x-linked, and results in no male live births, and a 50% reduction in female live births (Pliss et al., 2004). Biochemical results showed a reduction in brain and liver pyruvate dehydrogenase activity by about 25%. Cells showed a decrease in immunostaining for pyruvate dehydrogenase in the neocortex, cerebellum, and basal ganglia. The caudate and putamen showed a decreased organization of fibers as compared to controls. The cerebellum was overall smaller, with granular and Purkinje cells being reduced in numbers. Midbrain structures were largely unaffected.

This model of pyruvate dehydrogenase deficiency allows the assessment of the deficiency on the developing CNS. Results showed that a threshold exists for the activity of pyruvate dehydrogenase such that a loss of more than about 25% is not compatible with life. This is in keeping with earlier studies in adult mammals showing that a drop of about 25% of activity in pyruvate dehydrogenase was associated with severe neurological symptoms and death (McCandless, D. and Schenker, 1968).

The question of selective vulnerability in thiamine deficiency has been one of the key unanswered problems and has been recently examined. Selective vulnerability means highly localized brain regions, and/or individual cell types are affected. This question arises because of the striking localization in human cases of Wernicke's disease of lesions to sites such as the mammillary bodies. Since the highly localized nature of the lesions is at the lower limits of direct biochemical analysis, alternative techniques have been utilized.

In one such study (Pannunzio, P. et al., 2000), cerebellar granule cells (CGCs) were utilized in culture to examine various aspects of thiamine deficiency in a specific cell type. Briefly, CGCs were prepared from 7-day-old rat cerebella, and cultured cells were given Dulbecco's modified Engle's medium, deficient in thiamine. Pyrithiamine was also utilized to produce thiamine deficiency.

Results showed that CGCs exposed to the thiamine-deficient media resulted in a 97% decrease in thiamine, whereas pyrithiamine treatment yielded a less dramatic decrease of 63%. Transketolase activity was decreased by exposure to thiamine-deficient media and to media containing pyrithiamine. In addition, alpha-ketoglutarate dehydrogenase activity was decreased by pyrithiamine treatment, whereas thiamine-deficient media alone did not affect alpha-ketoglutarate dehydrogenase activity. Pyruvate dehydrogenase activity was not changed by either treatment. In addition, ATP levels were lowered by 35 and 68% with deficient media and media plus pyrithiamine, respectively. Pyrithiamine treatment resulted in a 27% increase in cell death.

This study represents the first to report effects of thiamine deficiency on cultured cerebellar cells. The decrease in alpha-ketoglutarate dehydrogenase activity was greater in pyrithiamine-treated cells than was the decrease in transketolase. Thiamine-deficient media alone did not alter enzyme activity levels, although thiamine biphosphate levels were lower. That the 80% decrease in alpha-ketoglutarate dehydrogenase activity in pyrithiamine-treated cells led to a drop in ATP is not surprising. This drop in enzyme activity is greater than that seen in vivo. In addition, a lowering of pyruvate dehydrogenase activity is seen in vivo, but was not observed in these cell culture experiments.

In another study (Ke, Z. et al., 2003), the time course and length of time before irreversible effects occurred were determined. The authors determined that by days 10–11 after initiation of thiamine deficiency via dietary means plus pyrithiamine injections, lesions occurred that were essentially not reversible.

Results showed the most sensitive brain area was the thalamus, specifically the submedial thalamic nucleus. Clinically, the mice were reversible up to days 10–11, at a time when symptoms were severe. By day 11, the altered histological picture included the entire thalamus. Symptoms could be reversed up to days 10–11. Neuropathological changes included hemorrhagic necrosis and degeneration of neurons. This started in the submedial thalamic nucleus and then spread to the other thalamic nuclei. Hemoxygenase-1 microglia were increased by day 10 of treatment, while NeuN-positive neurons were decreasing in numbers.

Results also showed that after 8 days on the thiamine diet with pyrithiamine, thiamine administration could not reverse impending cell death but prevented further cell death. Microglial responses were similar to those of neurons. Some irreversible changes had occurred by day 9, but treatment with thiamines slowed or stopped the process.

The authors point out that studies of neuronal cell death need to be correlated with the responses of other cells such as microglia, astrocytes, endothelial cells, and mast cells. The authors suggest that in this model, somewhere in day 9, the tide turns, as it were, and irrevocable changes cannot be avoided. Administration of thiamine on day 11 produces an increase in the numbers of microglia and HO-1 positive microglia, thereby increasing phagocytic activity. That there are clear differences between thiamine deficiency in man and the rapid production of thiamine deficiency in animals using pyrithiamine should be remembered.

In an attempt to localize and examine mechanisms of the distribution of lesions in thiamine-deficient rats, histamine positive neurons and localization of mast cells in the inferior colliculus and the thalamus were studied (Meng, J. and Okeda, R., 2003). Immunohistochemistry was used to determine the distribution of mast cells, and vulnerable and non-vulnerable regions of thiamine-deficient rats. Results showed similar results between vulnerable and non-vulnerable regions in thiamine-deficient rats. The authors concluded that a role for histamine in thiamine-deficient lesions could not be supported from these studies. The authors speculate that spongy lesions seen in the inferior colliculus and the thalamus are due to ischemia. These spongy lesions are not seen in the cerebral cortex or other unaffected areas. The acute regional neuronyl changes may be due to the ischemic condition of the spongy areas. Proof of this awaits further experimental studies.

In another recent paper (Wang, X. et al., 2007), endoplasmic reticulum stress (ER stress) was examined in thiamine-deficient rats and mice. Tissue culture from the cerebellum was utilized; thiamine deficiency was produced by administering a thiamine-deficient diet supplemented by daily injections of pyrithiamine.

The effect of thiamine deficiency was examined using various methods. For example, there was an increase in GRP78 expression in thiamine-deficient mice, reaching a 5.5-fold increase on day 7 of treatment. The expression of GADD153 and phosphorylated eLF2 alpha correlated with the expression of GRP78. Caspase-12 activation was also deemed increased by 4.6 times on day 6 of thiamine deficiency. The neurological status of thiamine-deficient animals was not stated, and pair-fed controls were not used.

The thalamus was also subjected to analysis since that brain area is affected by thiamine deficiency. Results were similar to those described in the cerebellum, except that caspase-12 was unchanged.

A thiamine analog, amprolium, was also used to evaluate cerebral changes. Results showed that amprolium-induced alterations in markers of ER stress similar to changes noted with a low-thiamine diet plus pyrithiamine. Caspase-12 was also increased.

The authors speculate that these data support the idea that ER stress may be tied to CNS lesions. Other CNS lesions such as those of Alzheimer's disease and Huntington's disease are associated with changes suggestive of ER stress. The authors further suggest that since thiamine deficiency and other neurodegenerative disorders share some common features, thiamine deficiency might serve as a model for these other disorders. The authors state that the change in ER stress occurs on day 5 after inanition of thiamine deficiency, therefore preceding the lesions. Whether these changes in ER stress actually are directly responsible for the neuropathological changes in brains of thiamine-deficient animals remains unclear. The changes in caspase-12 in lesions of thiamine-deficient animals can be blocked by Z-ATAD-FMK, a synthetic peptide that inhibits caspase-12. This inhibition acts to protect against thiamine deficiency induced cell death. The mechanism of this finding is also unclear. Remember that this study was performed using the antithiamine compound pyrithiamine, which produces thiamine deficiency very quickly (three times faster than thiamine-deficient diet alone), and without the benefit of pair-fed controls, which account for possible effects of starvation. Starvation alone produces significant alterations in brain chemistry.

Beriberi

The condition known as beriberi has been recognized for hundreds of years. The first written description was published in 1642 by Jacobus Bontius. Bontius was a Dutch physician who traveled to the West Indies and observed several tropical disorders including beriberi. These observations were published in a book entitled de Medicina Indorv Lib 4. A translation of beriberi observations appears in Appendix B. Interestingly, a couple of years after the Bontius book appeared, another Dutch physician, Tulp, also published a description of beriberi. Tulp was also well known as the subject of a famous painting by Rembrandt titled "Dr Tulp's Anatomy Lesion."

It is important to note that beriberi is largely a disease of the past. In the chapter on thiamine deficiency and world health, various outbreaks of thiamine deficiency and beriberi are described, but these are isolated occurrences. Beriberi was a major health problem 100 years ago in areas such as Malaya. In areas where wages were low, workers consumed large quantities of polished rice. Chinese mine workers, for example, working in jungle locations relied on food brought in. This usually consisted of large amounts of polished rice. Frequently all daily meals consisted mainly of rice. It was estimated that a worker might consume as much as 2 lbs (!) of rice per day. The sad outcome was that in 1900, one estimate was that in a remote mine, 800 of 2400 workers died from beriberi in a 2-year period.

As time went on, roads were built to remote mining areas, facilitating shipment of food. Mining areas became towns, with increased demand for food variety. This brought a decrease in the incidence of beriberi. Even reported beriberi incidences are low due to the fact that many cases are of a mild chronic neuropathy, which was unreported. Another interesting phenomenon was that with prosperity came increased cases of beriberi. The reason was that people could purchase highly polished rice, whereas in bad times, they polished their own with less effective means. There are many reported cases in which the incidence of beriberi rose following the introduction of small rice mills into a community. WW2 certainly brought about conditions in which people consumed what was available, increasing rice consumption and the incidence of beriberi.

Beriberi in male adults usually falls into one of three categories: a chronic dry atrophic type, a mild subacute form, or an acute fulminating type. The subacute mild

D.W. McCandless, *Thiamine Deficiency and Associated Clinical Disorders*,
Contemporary Clinical Neuroscience, DOI 10.1007/978-1-60761-311-4_4,

form of beriberi constitutes about 80–90% of all cases. The first type of beriberi is usually seen in older patients. This form is also associated with alcohol consumption. In these cases, wine is frequently warmed in lead-lined pots, and lead intoxication is part of the condition. There is also an association with opium addiction (Platt, B. and Alley, R., 1942). The acute phase of beriberi can convert to the dry chronic form in certain circumstances (Lim, E., 1938).

Clinical manifestations of the mild subacute cases of beriberi vary somewhat. The signs and symptoms of cardiovascular involvement in this form of beriberi are more significant than any others. They generally consist of breathlessness on any activity, and of chest palpations. There is edema present in most cases, and it may be associated with congestive heart failure. The heart is almost always enlarged, and tachycardia is present. Cardiac dilation may be the first cardiovascular change and usually precedes neurological changes (Aalsmeer, W. and Wenckebach, K., 1938) (Fig. 1).

Beriberi Clinical Symptoms

Source	Breathlessness	Chest Palpations	Edema	Enlarged Heart	Tachycardia	Neuropathy
Aalsmer, Et.al.	+	+	+	+	+	+
De Langen & Lichtenstein	+	+		+	+	
Denny-Brown		+	+	+	+	+

Fig. 1

Frequently, the onset of symptoms is linked to a febrile episode. This is associated with an increased need for thiamine, documented many times elsewhere. Patients in the subacute group may also show a deficiency in other essential nutritional substances such as ascorbic acid, niacin, and vitamin A. Most of these patients also suffered from protein deficiency, evidenced by leg ulcers of a type associated with protein deficiency (Platt, B., 1953).

In one situation, in an area where there were several factories that provided meals, workers from one factory had a much lower incidence of beriberi than others. The difference was that the beriberi-free factory served rice which was milled, washed, and cooked in a way such that almost no thiamine was lost. Other groups used commercial sources of rice prepared in a way that lost almost all of the thiamine. In the factories using commercial rice, over time almost 75% of workers developed edema and early signs of thiamine deficiency, whereas workers at the site carefully preparing their rice daily had only a couple of cases of edema. Workers in factories fed commercial rice were apathetic and listless, while workers fed daily prepared rice that maintained thiamine content were "high spirited."

The acute fulminating form of beriberi has an incidence of only about 5% of all beriberi cases. This form of beriberi is the most serious form, and until thiamine treatment was developed and utilized, nearly all patients died. The signs and symptoms of patients with the severe form have been described (de Langen, C. and Lichtenstein, A., 1938). This clinical account has been cited as an excellent description by others (Hawes, R. et al., 1937) and is now quoted verbatim from the above reference (de Langen):

"The whole picture here is dominated by the insufficiency of the heart and vessels, and usually runs a fatal course. The milder attacks may pass over to this type, but as a rule pernicious beriberi breaks out suddenly in a peracute form, before any other manifest symptom has proclaimed the presence of beriberi in the patient. When the nerve lesions develop early, we do not see Shoshin; the heart is saved from this extreme insufficiency the patient's being forced to a complete rest at an early stage of his attack. If the nerve symptoms come on late, and the patient with his labile heart and vessels can remain for some time on his feet, and must often even continue to work in order to earn his daily bread, the best possible conditions are present for the sudden development of Shoshin. Patients are sometimes known to die of this form of beriberi when next to no edema are as yet demonstrable.

The death from Shoshin beriberi is often very terrible. The patients are severely dyspnoeic. Violent palpitations of the heart allow them no moment of rest. An intense precordial agony is often one of the most distressing symptoms. The patient is very restless; he cannot lie still for an instant, but tosses and turns about violently from side to side in bed the whole time. The precordial oppression is often felt in the epigastrium. It is a feeling of heaviness, constriction, and oppression, sometimes of actual pain. The patient moans and shrieks, and his cries take on a rather special character because of the coincident hoarseness or sometimes aphonia. He is often intensely thirsty, but reacts to drinking by vomiting. He has an anxious look on his face. His pupils are dilated; his respirations frequent and superficial.

In spite of his dyspnoea his accessory muscles of respiration are brought but little into action, often not at all. He does not want to sit up in bed and help his respirations with both hands as is the case with other dyspnoeic patients, but, so far as his restlessness will permit it, he prefers to lie flat in bed. He does not cough or spit. In the heart region, the epigastrium and the neck, widespread and powerfully undulating pulsations are to be seen. There is an obvious cyanosis, more marked during inspiration. The pulse is fuller than might be expected, is usually regular, even and weak, with a frequency of from 120 to 150 per minute. A very high frequency of the heart is seldom seen. Over the heart area a wave-like motion may be felt. The liver is enlarged and tender. The epigastric region is spontaneously painful. Percussion shows the heart to be enlarged in all directions, but mostly to the right. The broadening of the area of aortic dullness, caused by the vascular stem, is plainly demonstrable extending into the left. Changes in position affect the heart outline markedly; it conforms to the position assumed by the patient. Over the body of the organ systolic murmurs may sometimes be heard, but not always. The heart action may only be described as tumultuous, although the sounds are not so powerful as in the last type of the disease described. Murmurs and sounds are difficult to localize. Vessel sounds are heard not only over the femoral artery, but also over the vessels of the elbow. These may become so loud that, as Shimazono has reported, they may even be heard without the aid of a stethoscope and in some instances at a distance from the bed. The systolic pressure, which at first remains normal, finally falls slowly to below 100, often reaching 80 or even 70.

Strangely enough there is but little disturbance to be found in the lungs; at most there is a mild typany on percussion. Just before death there tends to be a slight dullness at the lower borders. The respiration is rough and whistling, but without rales. As the condition becomes worse, rales may appear, at first dry and high-pitched, but soon becoming more moist. At the same time the pulse becomes thinner, the veins dilate more and more and the patient dies intensly dyspnoeic, but usually fully conscious. It is but seldom that a patient survives an attack, but even if we happen to be fortunate enough to bring him into a less immediately threatening condition, the risk of a sudden relapse is never absent."

There are some differences in symptoms in the severe form, one of which is the degree of edema shown by various patients. Some patients have little to no edema, while others have a great deal. Since classification of the type of beriberi between wet or dry depends in large part on the degree of edema, the absence or near absence of edema can make classification difficult. It should be remembered that the relative amount of edema in any given patient depends on conditions such as vomiting, heart failure, renal failure, and possible alcohol consumption.

Pyruvic acid is elevated in blood of patients with the severe fulminating variety of beriberi and is also elevated to a lesser extent in the mild form. Upon administration of thiamine, pyruvic acid returned to normal values in only a few hours. At the same time, there was an overall improvement in signs and symptoms, emphasizing the "biochemical lesion" nature of the disorder. Other changes reverting toward normal occurring with thiamine treatment, such as heart dilation, take several weeks. Also, diuresis in patients with severe edema may take several days to begin and several more days for edema to subside (Fig. 2).

Beriberi: Frequent blood findings

- Increased pyruvic acid
- Increased lactic acid
- Increased blood volume
- Decreased thiamine
- Increased thiamine pyrophosphate effect
- Decreased magnesium

Fig. 2

Most knowledge of the pathological changes seen in patients dying of beriberi are derived from early studies (Follis, R., 1958a). Much of this is because, since the development of thiamine knowledge, and treatment programs, which are widespread and simple, death from beriberi is rare. At a time when autopsies were being done on beriberi patients, descriptions and techniques were somewhat primitive.

Cardiopathology at autopsy is usually described as a dilation of the heart with concomitant hypertrophy (Follis, R., 1958b). Microscopically, fragmentation of cardiac fibers and fatty infiltration was seen. Also noted was cloudy swelling and areas of frank necrosis. Edematous and hydropic changes in cardiac muscle have also been noted. No areas of inflammation have been described. These pathological changes may be only moderate because this is a biochemical lesion, and more structural changes could be expected over a longer time course.

Changes in peripheral nerves consist of myelin loss and nerve fiber axoplasm disruption. Degeneration in the spinal cord was described (Durck, H., 1958), although minimal changes in the brain were described. Sometimes changes in medullary neurons were noted, but certainly none of the dramatic changes seen in patients with Wernicke's disease.

Beriberi has been described from a neurological point of view in prisoners of war (Denny-Brown, D., 1947). In this description, the key feature in prisoners with beriberi was the presence of foot and wrist drop. This finding was associated with tenderness of muscles over the lateral portion of the legs, lateral aspect of the thighs, and lateral forearms. Also seen were pain in the foot, paresthesias, and absence of ankle jerks. Ataxia was not an early sign of beriberi.

It is noted that alcoholic polyneuropathy could be quite similar to that seen in beriberi. A key difference, however, is that in the case of alcohol-related symptoms, treatment of the patient with thiamine while continuing alcohol intake improves the symptoms. Also, in alcoholism, there is usually more widespread sensory neuropathy and burning paresthesias in the feet.

Cardiac changes were seen in essentially all prisoners suffering from beriberi. Tachycardia was usually present, as well as palpitations. Also present is congestion in the venous circulation. Warm climates, and/or fever can precipitate more overt symptoms in chronic patients. The seasonal association of onset of symptoms in warm climates acts to increase congestion in venous circulation, precipitating cardiac pathology instead of neurological complications. The edema is of course a concomitant of the peripheral circulatory condition. This might explain why thiamine deficiency in the orient produces beriberi, whereas thiamine deficiency in cooler climates (US) is more conducive to the production of a neurological disorder (Wernicke's disease). In keeping with the above, elevated pyruvic acid levels in blood are usually present in beriberi in warm climates, but elevated pyruvic acid in cooler climates and in Wernicke's disease is not so constant.

While beriberi is usually thought of as an adult disease, more descriptions recognize a form of beriberi that affects infants. The infantile form of beriberi has been described by Burgess (1958). He states that there were incidence rates which varied between countries. For example, in Japan, only 3.5 deaths per 1000 births could be attributed to beriberi. In the Phillippines, the incidence of beriberi in infants under 1 year at autopsy was 56%. Also noted was that nearly all deaths occurred between 1 and 3 months of age and that these infants were breast fed. Andrews (Andrews, V., 1912) demonstrated the connection to breast milk by feeding newborn puppies human breast milk from mothers with beriberi, and thus producing beriberi in the newborn dogs. Around 1912, investigators (Chamberlain, W. et al., 1912) were able to use an extract from rice polishings, which was able to treat mild cases of beriberi and could be used to prevent the disorder. By the early twentieth century (Albert, J., 1947), extensive use of the extract from rice polishings had cut the mortality rate of infantile beriberi by 80%. By the end of WW2, the incidence had further declined.

Other countries reported outbreaks of beriberi. There was an outbreak in the Nauru Island (Bray, G., 1928); see also the World Health chapter in this volume. The age of onset was from the 8th to 12th week of age, and the course was rapid. Many infants died, but the mothers showed little beriberi symptoms. Another location of infantile outbreaks was Singapore where hundreds of cases were seen. In 1945 over 1000 cases of infantile beriberi were described in Madras (Krishnan, B. et al., 1945). In Hong Kong there was a large incidence of infantile beriberi, accounting for a mortality rate of 300/1000. At least 18% of newborns coming to clinics were affected, and the mortality rate was 90% (Fehily, L., 1944). There are many more reports of similar outbreaks. Many other outbreaks from isolated areas are probably under-reported or not reported at all.

One reason given for the decline in the incidence of infantile beriberi after WW2 is the increased status of women. As their status improved, they were able to eat all of the foods prepared for meals instead of leftovers. In some countries, there was a habit of restricting the diet of mothers just having given birth. The restriction to rice and dried fish lasted around 6 weeks, putting the mother at risk for thiamine deficiency. Once this practice was reduced, infant mortality from beriberi dropped.

It is noted that the mother of infants with beriberi frequently have signs and symptoms of beriberi. This finding is associated with the acquisition of infantile beriberi by the offspring. In these cases, breast milk seems to be the direct cause of morbidity and mortality, whereas in most cultures breast milk promotes healthy babies.

The onset of infantile beriberi between the first and fourth postnatal months probably reflects some storage of thiamine in the newborn due to placental transfer of the vitamin, coupled with lower requirements for B1 immediately after birth. Onset after 4 months of age usually signals a more chronic course.

Clinically, onset is rapid. The first signs may be a sudden attack of screaming, with stressful respiration and cyanosis. This phase passes, but recurs with increased frequency and intensity. Unless thiamine treatment is quickly initiated, death ensues. Aphonia may accompany this acute attack in which the child appears to be crying/screaming, but no sound is heard. This is attributed to laryngeal nerve paralysis. A variation of infantile beriberi is one in which some features of Wernicke's disease are seen, including nystagmus, strabismus, seizures, and an elevation of body temperatures.

Pathological studies of infantile beriberi are spotty, as they are in adult beriberi. The heart in infantile beriberi is described as being greatly enlarged – as large as that of a 10-year-old child. The pulmonary artery and aorta are equal in size, and microscopically, muscle fibers appear swollen. Congestion and edema have been described in liver, spleen, and brain. Biochemical tests of maternal breast milk show a lower thiamine level in areas in which infant beriberi is common as compared to breast milk in areas where the disorder is non-existent (Ramasatri, B. and Indravathi, D., 1957). Levels in areas of infantile beriberi were one half those of healthy areas. The volume of milk produced by mothers of infantile beriberi infants seems to be about the same as healthy mothers.

Treatment of infantile beriberi with thiamine may have obvious signs of improvement in as little as 6 hours and possibly earlier. This is clearly a biochemical lesion, and certainly pure thiamine deficiency. Such rapid return of function has repeatedly been seen in experimental animals with thiamine deficiency.

Reducing and/or eliminating the incidence of beriberi both in adults and in infants involve increasing the intake of thiamine. One way to achieve this is to improve the diet such that rice becomes a much less important part of the diet. This goal is being achieved around the world. A second way is to decrease the amount of milling of rice, which increases the thiamine content of the rice. This too is decreasing as knowledge of the dangers of highly milled rice increases. Washing is another procedure in the preparation of rice that decreases thiamine content significantly. Another method of increasing thiamine content of rice are enrichment programs. The actual enrichment process is relatively simple, but there are many small mills throughout affected areas, and it is not practical to oversee the preparation (and enrichment) of these widespread mills. All of these various programs have been responsible in part for the highly significant decrease in beriberi world wide. Finally, when beriberi does occur, modern medical abilities for rapid diagnosis and treatment have lowered the mortality rate.

As regards animal models of beriberi (see also the chapter on animals models of B1 deficiency in this book), a large number of studies have been published. When a thiamine-free diet is fed to mice and rats, the main symptom complex relates to the nervous system. Two major thiamine antagonists have been used: pyrithiamine and oxythiamine. Of these two, pyrithiamine primarily has an effect on the nervous system, whereas oxythiamine not only decreases the excitability of cardiac cells and increases the duration of the action potential but also reduces Purkinje fiber conduction velocity and spontaneous beat.

Experiments using oxythiamine in conjunction with a thiamine-free diet have been conducted in mice, rats, and pigeons. Polyneuritic symptoms in experimental animals on oxythiamine are rarely seen; however, bradycardia is easily produced in rats (Gurtner, H., 1961).

Results from oxythiamine administration include a decrease in conversion of pyruvate to acetoin, and a lowered transketolase activity (Brin, M., 1962). Another study showed that oxidation of pyruvate was decreased in rat heart in oxythiamine-induced thiamine deficiency, whereas oxidation of alpha-ketoglutarate activity was decreased only in heart tissue (Gubler, C., 1961). Blood levels of pyruvate are also elevated in animals rendered thiamine deficient with oxythiamine.

The main oxythiamine-based antagonist is not oxythiamine itself, but also the phosphate ester, oxythiamine pyrophosphate (OTPP) (Koedem, J. and Steyn-Parve, E., 1960). OTPP is a strong inhibitor of transketolase and thiamine pyrophosphate. OTPP has a greater affinity for transketolase than does thiamine pyrophosphate, thereby displacing it from the haloenzyme. Therefore, the mode of action of oxythiamine is as a competitive inhibitor for sites on the haloenzyme. It successfully competes with thiamine pyrophosphate for these sites, then cannot function as a coenzyme, thus lowering enzyme activity. The advantage of using antimetabolites in experimental animals is that they speed up the development of symptoms. The two antimetabolites pyrithiamine and oxithiamine allow the "splitting" of the symptoms into neurological (pyrithiamine) and cardiac (oxythiamine). Thus, while their use has not uncovered any particular phenomenon not already known, they have emphasized changes in brain and in heart by separating the two symptom complexes.

In man, clinical symptoms in beriberi consist of those of cardiac and neuropathological origin. Beriberi heart disease consists of exertional dyspnia, tachycardia, and heart failure. This progression can occur very quickly, but can be reversed by thiamine administration. This condition may be accompanied by edema, a usual hallmark of beriberi heart disease. Key symptoms of Wernicke's disease such as nystagmus, opthalmoplegia, and amnesis are almost never seen in beriberi heart disease.

In an early study of cardiac lesions in thiamine-deficient rats (Ashburn, L. and Lowry, J., 1942), thiamine deficiency was produced by dietary means, and the hearts were examined for pathology. Results showed several rats had a clear fluid in the chest cavity. Hearts were enlarged especially in the right auricle. The left auricle was also dilated, but to a lesser extent. The wall thickness appeared normal. Microscopically, heart muscle fibers were hyalinized, and showed increased oxyphilia in the cytoplasm, and the nuclei were pyknotic or even absent. Necrotic muscle fibers were noted; some fibers were fragmented. Interstitial proliferation was seen, and muscle fibers in these areas were usually absent. There was a significant proliferation of fibroblasts. In its worst manifestation, the entire wall thickness of the chambers was composed of fibroblasts. Thrombi in the auricular wall were occasionally seen. From this sampling method, all parts of the auricles were affected by thiamine deficiency. Examination of the ventricles showed much less frequency of lesions. Out of 57 hearts examined, only 9 showed any ventricular involvement. When present, ventricular lesions were smaller than those in the auricles. It was also concluded based on the time course of the development of symptoms and lesions that cardiac pathology is a feature of the late stages of dietary thiamine deficiency in the rat (Fig. 3).

In a study of thiamine deficiency in dogs, it was shown that myocardial utilization of pyruvate and lactate was abnormal (Hackel, D. et al., 1953). In fact it was found that the extraction of pyruvate by myocardial tissue was decreased, while the utilization was increased. The coefficient of extraction of lactate was also lower, as was lactate utilization. The utilization of lactate was greater than the utilization of pyruvate. These data suggest defects in myocardial metabolism of pyruvate and lactate, and in these studies, starvation was not a factor since animals were eating well and had only minor weight loss.

In another study of thiamine deficiency in rats, EKGs were performed on normal controls and deficient rats (Yoshitoshi, Y. et al., 1961). The authors state that there are many variations in reported findings from EKGs in thiamine deficiency. Findings in this study included arrhythmias in 17% of cases in later stages. The heart weight increased from the second week through the fifth week. After day 16, fatty infiltration appeared in the myocardium representing a disturbance in glycolysis. Changes in ST and T in the EKG were due to myocardial lesions after the fifth week on the deficient diet. These ST and T changes consisted of both elevation and depression. Other cardiac changes included an increase in right axis deviation, A-V block, premature beat, and interpolated nodal rhythm.

In another study of thiamine deficiency, effects on sarcosomes' oxidative phosphorylation and electron transport were described (Arcos, J. et al., 1964). Results showed an increase in the heart weight–body weight ratio. Oxidative rates of cardiac tissue with pyruvate and alpha-ketoglutarate showed a faster depletion of thiamine pyrophosphate with pyruvate as substrate than alpha-ketoglutarate. With malate as substrate, respiratory rates were barely changed from controls. In contrast to respiratory rates, P:O ratios were unchanged. ATP values (not directly measured) were probably depleted as indicated by the inhibition of sarcosomal swelling by

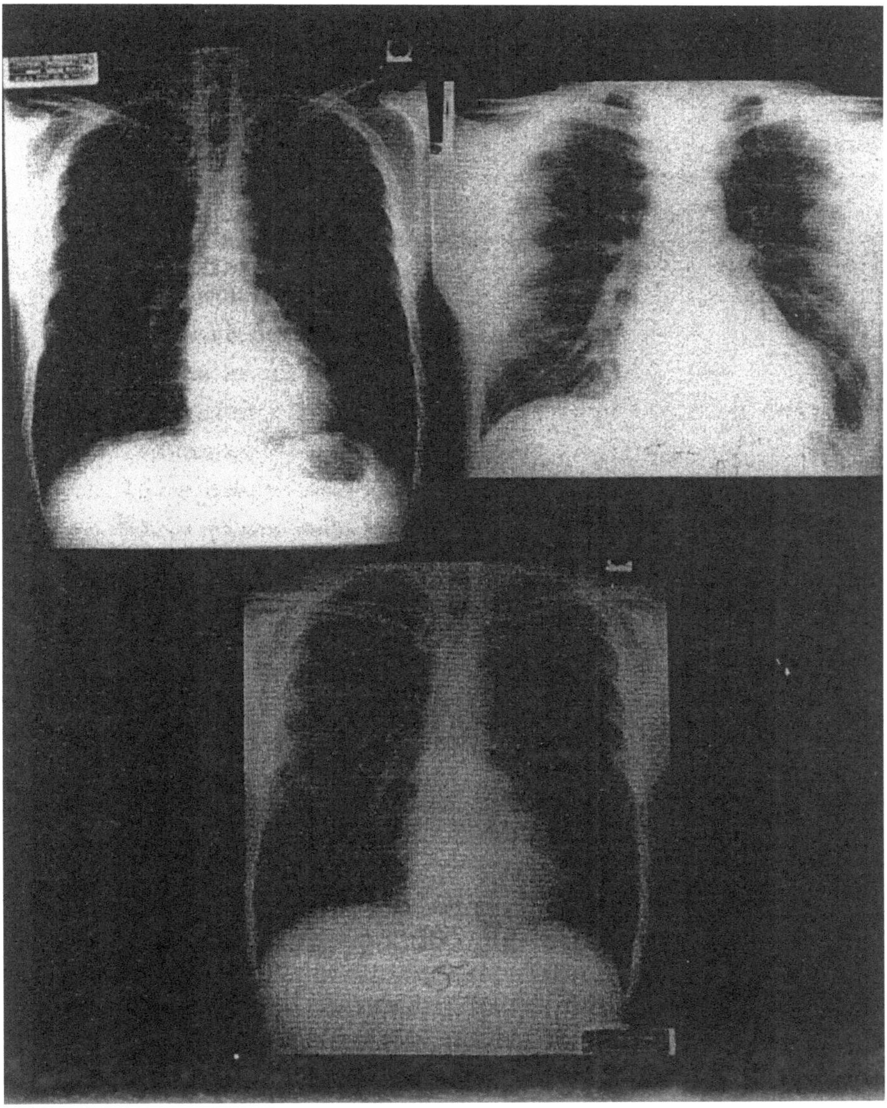

Fig. 3 Chest X-rays: *top left* = before beriberi, *top right* = during acute beriberi, *bottom* = after 11 days of thiamine treatment. Reproduced from Attas et al. (1978) with permission from Wolters Kluwer

pyruvate–fumarate substrates. Decreased inhibition of swelling shows a decreased ability of sarcosomes to generate ATP from pyruvate metabolism.

Since the myocardial isometric tension is an indicator of working capacity of muscle, a study was performed to determine the effects of thiamine deficiency in rats on this physiological phenomenon (Aldinger, E., 1965). Thiamine deficiency was produced by dietary means. The measured myocardial tension in thiamine-deficient rats was only a small fraction (25%) of that in control rats. For the first 2 weeks, however, contractility was no different than controls, only beginning to drop at 3 weeks. The author also noted arrhythmias and ventricular fibrillation in a significant number of rats. Some rats on the diet for 4 weeks, which were administered thiamine as a treatment, returned to normal physiologically in only a few hours, whereas rats on diet for as long as 5 weeks (and terminally sick) did not recover with thiamine treatment.

The results of this study are in keeping with previous studies in that by 5 weeks on diet, when animals are severely affected, thiamine treatment is to no avail. This finding suggests structural damage. The author implies his findings are consistent with the hypothesis that the thiamine deficiency produces lowered levels of thiamine pyrophosphate, thereby decreasing the potential for oxidative metabolism and energy production. This would in turn lead to the findings described in this paper on myocardial contractibility.

In a study using electron microscopy to examine thiamine-deficient hearts, Wistar rats were fed polished rice and compared to control rats fed the same diet plus thiamine (Suzuki, T., 1967). With this feeding regime severe thiamine deficiency took up to 8 weeks to develop.

Results showed that hearts harvested from thiamine-deficient rats on diet for 5–8 weeks were slightly dilated and slightly hypertrophied. Examination with light microscopy, some hearts showed fatty degeneration, disappearance of cross striations, some necrosis, and edema and cell infiltration. Examination using electron microscopy showed swollen mitochondria at 5 weeks on the thiamine-deficient diet. Some mitochondria showed destroyed internal structure, with cristae absent. By 8 weeks on diet, swelling of mitochondria was widespread. Most were irregular in shape, with cristae quite deranged. The sarcoplasmic reticulum was smaller and ruptured. The severity of these structural changes including the absence of cross striations at this time could easily explain the sudden death in the eighth week.

When the thiamine-deficient rats were "reversed" by thiamine administration these structural changes showed diminution and a return toward normal structure. Thus, in a group that were on a deficient diet, then treated with thiamine for 2 weeks, mitochondria were only slightly enlarged. The cristae had an arrangement approaching normal, but were still scanty. The sarcoplasmic reticulum was nearly normal in shape, and myofibril cross striations were distinguishable.

Following 4 weeks of treatment of 8-week deficient rats, mitochondria were only very mildly enlarged, and the cristae were all well preserved and densely arranged. The sarcoplasmic reticulum was not significantly different from controls. Myofibril cross striations could easily be seen. The author states that the finding of a few degenerated mitochondria in the hearts from the 8-week diet/4-week treatment group suggests that severely affected mitochondria were still not functional but that more mildly affected mitochondria had recovered function.

An interesting significance of these observations is the ability of thiamine to reverse not only the signs and symptoms, but also the structural changes.

Common thinking was (and is) that thiamine deficiency represents a true biochemical lesion, and if it lasts long enough, then irreversible structural changes occur. These reversible structural occurred at the microscopic level. A separate example is the Wernicke-Korsakoff disease in which changes seen in signs and symptoms in early Wernicke's disease can be dramatically reversed with thiamine. Conversely, by the time a patient reaches the Korsakoff psychosis stage, reversal with thiamine treatment is problematic. Perhaps the structural changes need to reach a greater evolution than early microscopic alterations. So here in the heart is clear evidence of structural changes being completely reversed with thiamine treatment. This probably represents the "highest" level at which structural changes can occur without permanent deficit.

In another study (McCandless, D. et al., 1970), cardiac metabolism was directly assessed in order to determine if thiamine deficiency was a factor in the cardiac signs and symptoms of beriberi heart disease. In this study, thiamine deficiency was produced in rats using dietary means. Controls consisted of normally fed control diet rats and rats pair-fed to the deficient rats. Deficient rats were fed thiamine-deficient diet only.

Results showed rats at 5 weeks with neurological signs had decreased transketolase activity in blood, and a higher TPP effect as compared to pair-fed controls. In these symptomatic rats, whole heart transketolase and pyruvate dehydrogenase activities were depressed by about 65% and 80%, respectively. The hearts of symptomatic rats showed an 18% hypertrophy. In addition, cardiac ATP was decreased by about 17% at 4 weeks in rats on the deficient diet. At 5 weeks the ATP decrease was 34%. The depletion of cardiac ATP correlated with both the fall in PDH activity, and with the rise of cardiac pyruvate. The drop in ATP returned to normal values within 24 hours after "reversal" with thiamine. The fall in ATP did not correlate with changes in transketolase. Both of these enzymes are thiamine (thiamine pyrophosphate) dependent. A decrease in ATP can result from decreased synthesis or increased utilization. The data suggest the ATP drop was due to a decrease in synthesis. The observed fall in cardiac ATP may produce an impairment in cardiac contractibility. Overt heart failure was not seen in spite of a 34% drop in ATP, possibly because many animals died from neurological complications. The ATP drop was about equal in atria and ventricles.

In a follow-up study (McCandless, D. et al., 1976a), the effect of thiamine deficiency on rat heart pentose phosphate pathway activity was assessed. Only animals displaying severe signs and symptoms of thiamine deficiency were used. The enzymes glucose-6-phosphate dehydrogenase (G-6-PDH) and 6-phosphoglucose dehydrogenase (6PGDH) were measured in pair-fed controls and thiamine-deficient rats. Results showed both G-6-PDH and 6-PGDH were not changed in deficient as compared to pair-fed control rats. Flux through the pentose phosphate pathway was estimated using the methodology of Katz (Katz, J. et al., 1966). While transketolase is the rate-limiting step in the shunt, G-6-PDH and 6-PGDH mediate flux. In keeping with the finding that G-6-PDH and 6-PGDH activities were not changed, in vivo flux through the pentose phosphate pathway was not changed in deficient as compared to control rats.

These studies were performed at a time in the course when symptoms were severe, and transketolase levels in previous studies showed activities only a fraction of control values. Based then on the levels of G-6-PDH and 6-PGDH, and flux through the pathway, these in vivo data do not support the hypothesis that flow through the pentose phosphate pathway is an important feature of the cardiac pathology seen in severe thiamine deficiency. Similar findings have been noted by us in regarding flux through the pentose phosphate pathway in liver and brain.

In another paper (Hauschildt, S. et al., 1973), the effect on isolated perfused rat heart of thiamine deficiency was analyzed. Rats were made thiamine-deficient via dietary means. After 21–28 days on the diet, at a time when neurological symptoms were severe, animals were decapitated, and the heart isolated and perfused with Krebs-Henseleit buffer. In this preparation, contractile force was measured in thiamine-deficient rats and controls. Lactate, pyruvate, ATP, ADP, AMP, and phosphocreatine were measured enzymatically. Results showed contractility strength in thiamine-deficient rats lasted about 15 minutes in this paradigm, compared to 2–3 hours in control rats.

As regards high-energy phosphates in this in vitro model, phosphocreatine and ATP were depleted, and ADP and AMP were elevated in thiamine-deficient rats as compared to controls. This was despite a high oxygen tension and adequate perfusion. When thiamine-deficient rats were reversed with thiamine treatment, high-energy phosphates and cardiac contractility returned to normal values. This again emphasizes the "biochemical lesion" nature of early thiamine deficiency. The authors state that two views are held regarding the pathogenesis of cardiac malfunction. The first states that thiamine depletion results in lowered activity of thiamine-requiring enzymes, with a resultant decrease in high-energy phosphates for cardiac work. The second hypothesis states that blockage of thiamine-dependent enzymes allows accumulation of a wide variety of metabolites associated with the affected pathway and that these accumulating metabolites serve as metabolic inhibitors and/or toxins for the affected organ. This in vitro study strongly suggests that the effect is direct, since there is good correlation between enzyme activities described in earlier studies and high-energy phosphates, and the correlation continues in the recovery phase. Certainly, in the present study the good correlation of lactate, pyruvate, citrate, alpha-ketoglutarate, isocitrate, ATP, ADP, AMP, and phosphocreatine lends strong credence to the direct effect hypothesis.

In a study (Takahashi, K. and Nakamura, H., 1976), nine cases of typical beriberi were studied as regards both cardiac involvement and peripheral neuropathy. The study emphasizes the fact that in classical beriberi cases, there is a significant neuropathological component.

These patients had eaten a polished rice diet with very little vitamin B1 supplementation. Patients may have presented acutely or subacutely. All showed edema, apical heart murmurs, accentuated pulmonary sounds, and cardiac enlargement. Blood lactate levels were elevated, while pyruvate levels were normal. Blood thiamine levels were below that of controls. Following treatment with thiamine, blood values, cardiac pathology, edema, and neuropathy returned toward normal in a few weeks.

During the pretreatment phase, sural nerve biopsy was performed in some patients, and in some other patients, biopsies were performed after treatment was begun.

Results showed a loss of myelinated fibers and an increase in collagen fibers were noted in the sural nerve. Fragmented myelin forms were seen in methylene blue stained sections. These changes were correlated with leg sensory loss. In electron microscopy sections, flattened vesicular forms were seen in axoplasm. Axonyl degeneration was noted in all cases examined. Also seen were degenerated mitochondria, vesicles, small myelin figures, and deranged Schwann cell cytoplasm. The "jelly role" appearance of myelin was frequently preserved suggesting the primary defect was axonyl degeneration.

In patients receiving thiamine treatment, many axons were seen via electron microscopy. These axons were not always surrounded by Schwann cytoplasm. No regenerating axons were seen in untreated patients' sural nerves. In two patients, signs of remyelination were evident, although 91 and 30 days elapsed between treatment and biopsy. The improvement of the scene observed with electron microscopy in treated patients correlated well with improvement in all signs and symptoms.

This study represents the first to look at neuropathy in well-documented beriberi patients. The light and electron microscopy changes in the sural nerves correlate well with similar changes in thiamine-deficient rats. Changes seen in the two patients showing remyelination were noted to be similar to other patients undergoing repeated myelination and remyelination. Large myelinated fibers seemed to be more affected in these patients than unmyelinated fibers. Since thiamine plays such an important role (thiamine pyrophosphate) in oxidative metabolism and energy formation, large fibers may be more sensitive to depletion of thiamine.

In another study, hemodynamic studies were carried out in four alcoholic patients with beriberi heart disease who did not have complicating cirrhosis, anemia, or other known heart disease (Akbarian, M., 1966). All four were chronic alcoholics with poor diet, and they all showed varying degrees of peripheral neuropathy. These four cases were well established by both clinical and hemodynamic data, and by the rapid response to thiamine treatment. All showed biventricular congestive heart failure and enlargement with pulmonary congestion. One patient showed atrial fibrillation and another patient had a complete left bundle branch block. These changes have been seen in other cases of beriberi heart disease and in thiamine-deficient rats. In all four cases, the presenting cardiac abnormality was high output failure. There was low vascular resistance with low left ventricular filling. The cardiac output was disproportionally elevated compared to the metabolic rate. Blood pressure was elevated in two of the four cases. There were associated high plasma and blood volumes in these patients. Clinically, the patients were confirmed as having left ventricular failure.

Treatment of the patients resulted in measurable improvement in only 37 minutes. In this case peripheral vascular tone and decreased left ventricular work resulted in low cardiac failure. Three patients received long-term thiamine treatment and had marked improvement clinically and physiologically. An increase in vascular resistance led to a decrease in heart rate, stroke volume, cardiac output, and oxygen consumption. The authors state that in addition to thiamine treatment, patients received overall good nutrition, bed rest, diuretics, etc. Patients are often treated as above, and controlled clinical studies looking at thiamine alone without bed rest, good food, etc., should be done in relation to beriberi heart disease. This type study has been done in terms of neuropathology in chronic alcoholics (see Wernicke's disease chapter).

It is noted by the authors that hard physical work acts to precipitate beriberi heart disease due to maintenance of a high cardiac work load. The peripheral neuritis so often seen in thiamine-deficient patients acts to prevent physical work, thereby lowering the incidence of beriberi heart disease in those patients. In keeping with that is the observation that adrenalin, which acts as a stimulator of cardiac output, can produce symptoms associated with beriberi heart disease (Shimazono, J., 1927). This was used as a "test" for beriberi disease in the clinical evaluation of patients suspected of beriberi. It is also noted that a tropical (hot) climate may exacerbate the development of beriberi. This explains the geographical location of high incidence of the disorder. The tropical climate is known to result in vasodilation and a high output state.

The strong link between the decarboxylation of pyruvate and cardiac muscle work has been demonstrated (Olson, R. and Piatnek, D., 1951). It was of interest to determine if the link between the decarboxylation of pyruvate and cardiac myofibrils was accomplished with or without the presence of mitochondria (DeArce, H. et al., 1966). In this study, glycerinated cardiac myofibrils were prepared from dog heart, separated from mitochondria, then incubated at various time intervals in the presence or absence of pyruvate, NAD, ATP, etc., and contractility assessed.

Results from these studies showed that the decarboxylation of pyruvate was linked to the contraction of myofibrils from the heart. This contraction occurred in the absence of mitochondria. The contraction was thiamine and calcium dependent. The separation of the carboxyl group from pyruvate provides energy that can be used for myofibril contraction. It appears that ATP is responsible for the tension aspect of cardiac myofibril contraction. It is possible that more than one source of energy is available for normal cardiac myofibril contraction. The close link of this process to thiamine availability explains how in beriberi heart disease, the absence of thiamine in the diet can lead to fatal cardiac pathology.

Early studies emphasized the vasculature aspect of beriberi, supposing that cardiac changes such as edema, fatty degeneration, and cloudy vacuolization of myofibrils, were secondary and were a result of increased work load and high output due to peripheral vascular pathology. However a case with clear evidence of beriberi heart disease was described with left ventricular pathology (Attas, M. et al., 1978).

This patient was a chronic alcoholic with significant cardiac signs and symptoms. Symptoms included a grade 4/6 systolic ejection murmur, and a grade 2/6 diastolic sound at the left sternal border. An EKG showed sinus tachycardia and non-specific T wave changes. Chest X-ray showed cardiac enlargement as compared to earlier chest X-rays. The patient soon showed a worsening biventricular failure. A diagnosis of severe acute (Shoshin type) beriberi was made, and 200 mg IV thiamine was administered. As usual, the patient responded rapidly to thiamine treatment.

The authors state that in this case there was a left cardiac pathology that was not secondary to vasculature alterations. They also state that the response to thiamine lags behind the vasculature response, leading to low cardiac output. The impairment of myocardial function could result from alterations including impaired cardiac energy metabolism, altered coronary blood flow, and/or the presence of other diseases known to have a cardiac component. Significant evidence for an alteration in cardiac energy metabolism has already been described (McCandless et al., 1970).

The case described is the only reported case of measured left ventricular function to date. Upon thiamine treatment, vascular resistance increased, with only trivial decrease in ventricular ejection. Therapy therefore resulted in a higher stroke work index, and a lower left ventricular end-diastolic pressure. This is seen as strong evidence for a depression in myofibril contractility during the acute phase of this patient's beriberi heart disease. This patient also developed severe lactic acidosis and an elevation of pyruvate. Elevation of these two metabolites is associated with beriberi. The control of lactic acidosis in this patient was complicated until thiamine treatment was initiated, emphasizing nutritional etiology of this disorder.

Not all cases of beriberi respond to thiamine treatment. In one case a chronic alcoholic was admitted to a hospital with dyspnea, paresthesia in the lower extremities, edema, ascites, and sensory disturbances (Betrosian, A. et al., 2004). The EKG was consistent with hypokinetic heart failure with ejection fraction less than 40%. Blood thiamine was less than half normal. By the third day the diagnosis of beriberi was made and thiamine was administered (100 mg IM/day). Some improvement was noted, but on the sixth day the patient expired. At autopsy the heart was small, and the myocardium was thin and atrophic. Microscopically, the heart revealed extensive fibrosis and edema, hyaline degeneration, and asbestosis. Peripheral nerves showed axonyl degeneration and myelin pathology.

The authors note that this is a case showing a "mixed" state, which included high output cardiac failure and polyneuropathy. The patient died despite prompt initiation of thiamine therapy. The authors speculate that the initial dose of thiamine might have met the needs of the peripheral nervous system, but not that of the cardiovascular system. The patient developed low cardiac output failure after treatment, which resulted in his demise. Clinicians should be conscious of the possibilities of this complication to thiamine deficiency and beriberi.

In another single case study, a 24-year-old pregnant patient with hyperemesis gravidarum was described (Indraccolo, U. et al., 2005). This patient claimed not to have consumed alcohol, but had been vomiting for at least 40 days and could eat only very small amounts of food. She was IV rehydrated and received IV glucose and antiemetics. She developed confusion, horizontal nystagmus, aphasia, and ataxia despite treatment. She lapsed into coma, and thiamine was administered. The diagnosis of Wernicke's disease was made in addition to the presence of wet beriberi. After 5 days of thiamine therapy the coma was gone, and the patient could partly feed herself. She continued to improve and gave birth to a normal baby.

This case of thiamine deficiency due to hyperemesis gravidarum had characteristics of both wet beriberi and dry (neuropathy) beriberi. Previous cases of hyperemesis gravidarum in pregnancy have been treated with thiamine doses ranging from 50 to 200 mg of thiamine per day. Doses over 400 mg/day IV are accompanied with risks such as nausia, anorexia, and ataxia. The pregnancy may be associated with an increased thiamine requirement. The dose used in this study, 300 mg of thiamine per day, then reduced to 200 mg/day when symptoms start to improve (fifth day), could prove beneficial in such cases.

In another case report (Tran, H., 2006), an elderly woman with chronic alcoholism was admitted with dyspnea. She had bilateral crepitations with chest auscultations, and a chest radiograph showed cardiomegaly and frank pulmonary edema. Blood thiamine was decreased 90% and transketolase activity also decreased. A diagnosis of wet beriberi with cardiac failure was made, and thiamine therapy was initiated. Thiamine treatment resulted in improvement of clinical signs in the following 2–3 days. The patient received regular thiamine supplementation and was discharged.

This case shows an under-diagnosed cause of congestive heart failure – wet beriberi and thiamine deficiency. Thiamine deficiency can be confirmed by blood thiamine determination and transketolase measurement, but these assays may not be available in many hospitals. Therefore, the diagnosis may have to be based on clinical suspicion, and the treatment empirical. In this case, thiamine treatment reversed the decreased left ventricular function.

Wernicke's Disease

The clinical entity Wernicke's disease was first described by Dr. Carl Wernicke in 1881. In this manuscript, Wernicke described both clinical features and autopsy findings (Wernicke, 1881). The description was of three patients, and clinical features consisted of ataxia, strabismus, nystagmus, diplopia, disorientation, and stupor. The findings at autopsy were similar and consisted of small hemorrhages located in periaqueductal gray. Wernicke interpreted this to be an acute inflammatory disease primarily of the ocular nuclei.

In 1887, Korsakoff published a paper describing a unique amnesic state that occurred in patients who were both alcoholic and non-alcoholic (Korsakoff, 1887). Korsakoff made the correlation between the alcoholic neuritis of Wernicke's disease, and the memory defect and confabulation of Korsakoff's psychosis. He considered that these were two facets of the same disorder (Korsakoff, 1889). Korsakoff also realized that these two features of the same disease might manifest in different proportions in different patients.

In another manuscript, Korsakoff (1889) describes the amnesia of this disorder as being of recent events, with a relative sparing of memory for more distant past events. Korsakoff notes that this type of amnesia occurs after prodromal agitation and confusion. The confusion may last several days, then resolves, but the amnesia remains. There exists a broad range of manifestations of these symptoms. Sometimes distant memories are also affected. Recent memories may not be clear, for example, a patient not being able to remember whether or not dinner had been taken. Another example is that a patient might say that he went to town yesterday, when in fact he was bedridden.

Pathological changes in Korsakoff's psychosis were first noted in 1896 (Gudden, et al., 1896). Brain lesions were described in the mammillary bodies, as well as in the periaqueductal gray. In a later neuropathological description (Gamper, 1928), the similarity between lesions in Wernicke's disease and Korsakoff's psychosis was noted. Another (Kant, 1932) investigator found brain lesions of Wernicke's disease in all of his fatal cases of Korsakoff's psychosis. Several subsequent studies described brain lesions which were very similar in both Wernicke's disease and Korsakoff's psychosis and included hemorrhagic changes in the periaqueductal gray, mammillary bodies, brain stem nuclei, and thalamus. These lesions could

D.W. McCandless, *Thiamine Deficiency and Associated Clinical Disorders*,
Contemporary Clinical Neuroscience, DOI 10.1007/978-1-60761-311-4_5,

explain the symptoms. It was Girard et al. (1956) who drew the conclusion that Wernicke's disease and Korsakoff's psychosis were similar in symptoms, neuropathology, and outcome. From this observation, the two disorders came to be called the Wernicke–Korsakoff syndrome. As time progressed reports of large (70 cases) studies of patients diagnosed with Wernicke's disease and/or Korsakoff's psychosis were published. These confirmed earlier descriptions of symptoms and location of neuropathological lesions. Studies of Korsakoff's psychosis patients described lesions identical in nature and location as those seen in Wernicke's disease. These authors all suggested that this represented variable clinical expression and outcome of the same disease.

It should be pointed out that there was some variation in different studies. Some authors, for example, found changes in neurons of the cerebral cortex, while others did not. Another example is that many neuropathologic studies did not mention cerebellar changes, whereas many recent studies have shown characteristic cerebellar changes (Figs. 1, 2 and 3).

Wernicke's Disease—Gross Brain Lesions

Source	Mammillary Bodies	Thalamus	Cerebellum	Cerebral Cortex	Periaqueductal Gray	Hypothalamus
Victor Adams & Collins	+	+	+	–	+	
Malamud & Skillicorn (Korsakoff's)	+	+	+	–		
Liu, et.al. (MRI Study)	+	+			+	
Jagadha, er.al. (non alcoholic Wernicke's)	+	+				+
Kornreich, et.al. (infant Wernicke's)	+	+			+	

Fig. 1

Wernicke's Disease—Microscopic Lesions

Source	Thalamus	Hypothalamus	Cerebellum	Mammillary Bodies	Pons	Medulla	Fornix
Victor, Adams, & Collins	+	+	+	+	+	+	+

Fig. 2

Specific neuropathological changes have been described in Wernicke's disease (Victor et al., 1989). These observations resulted from the neuropathological

Wernicke"s Disease—Type of Microscopic Lesions

Source	Cell Necrosis	Decreased Neurons	Increased Astrocytes	Hemorrhage	Loss of Myelin Fibers
Victor Adams & Collins	+	+	+	+	+
Malamud & Skillicorn	+	+	+	+	+
Jagadha, Et.al., (nonalcoholic Wernicke's)	+	+	+	+	+

Fig. 3

examination of 62 cases of Wernicke–Korsakoff syndrome over several years. In these studies, the surface features of the brains were normal, as was the weight.

Coronal sections were examined grossly and showed a symmetrical increase in the size of the lateral ventricles (26% of cases). The thalamus showed characteristic and symmetric areas of gray-brown discoloration in about 75% of cases. In addition, these lesions were described as spongy or granular in appearance. Hemorrhages were not frequently (less than 10%) seen grossly, and when present were usually in the mammillary bodies. Lesions were observed in the midbrain, as well as in the floor of the fourth ventricle, confined to the medulla. The cerebellum showed gross lesions in 36% of cases examined and was described by the authors as an atrophy of the folia of the vermis. The bilateral symmetry of the lesions was again noted (Fig. 4).

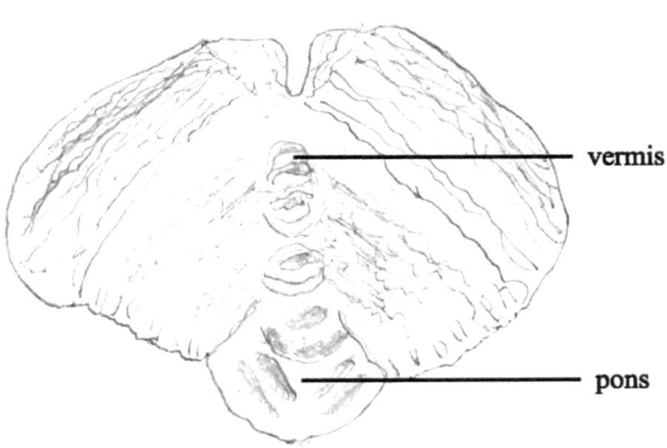

Fig. 4 Schematic cartoon at the level of the pons, showing the cerebellum

Microscopically, lesions from this study were noted to have a symmetric consistent location in the thalamus, hypothalamus, midbrain, pons, and medulla. Lesions were also noted in the cerebellum, as well as, to a lesser degree, in the cerebral cortex, hippocampus, and fornix.

The actual microscopic lesions showed variation due to the duration and course of the disease process. The most severe extent of the microscopic lesion was nearly complete cell necrosis in affected areas. Where lesions were not so severe both neurons and myelin were mostly absent. The most common degree of change consisted of some decrease in neurons, but many were spared. In this state, myelin was destroyed more than neurons. Also present was an increase in astrocytes and macrophages. Focal areas of hemorrhage were only occasionally seen. Blood vessel changes were a common and prominent change in affected areas, but were considered by these authors to be of a secondary nature. In some cases macrophages containing hemosiderin, indicating previous bleeding, were seen. These microscopic lesions were bilaterally symmetric.

Generally speaking, within certain damaged areas, such as the parafascicular nuclei of the thalamus, neurons were spared. There was a highly regional distribution of damage, for example, the dorsal motor nucleus of the vagus nerve was damaged in the upper medulla, whereas in the lower medulla, the vagus nucleus was unaffected. Most areas affected with microscopic lesions were gray matter. In all lesioned areas with damaged neurons, the myelinated fibers of the nerve cells seemed more affected than did the cell bodies.

Vascular and glial changes in lesioned areas were quite similar from case to case. The changes seen in neuropathology in the Wernicke–Korsakoff syndrome are different from those seen in either ischemia or edema. The relative sparing of neurons seemed to these authors to suggest a metabolic alteration. The exact reason underlying the unique specificity of the lesion pattern is unclear, but certainly this also occurs in Leigh's disease. The peripheral nerve changes in Wernicke's disease are also identical to those seen in beri-beri.

In another comprehensive study (Malamud and Skillicorn, 1956), the relationship between Wernicke's disease and Korsakoff's psychosis was studied. This was a clinical and neuropathological study of 70 patients in mental hospitals who had been diagnosed with Korsakoff's psychosis. The clinical progress was followed throughout hospitalization, and autopsies were performed upon death.

Autopsies revealed a variety of diseases such as cardiovascular, cirrhosis, carcinoma, etc., consistent with the mean age at death (61.6 years). Gross examination of the brains revealed significantly small mammillary bodies with a yellow-brown discoloration. In addition, the ventricular system showed dilatation in a small percent of cases.

The neurohistopathologic findings were revealing. The hypothalamic and thalamic periventricular areas had a diffuse gliosis in all cases using a Holzer stain. The mammillary bodies showed bilateral atrophy and gliosis in all but three cases. In most of these, the alteration was a chronic response of astrocytes and glial fibers. While hemorrhages were rare, there were frequently proliferative changes in the endothelium of blood vessels and an increase in reticular fibers. In some instances, there was a focal necrosis, with decreased numbers of neurons. The medial thalamus showed degenerative changes in the dorsomedial nuclei in over one half of the patients, where as the pulvinar was affected in only three cases. In these lesions, unlike those in the mammillary bodies, the neurons had nearly all disappeared and were replaced by glial fibers and astrocytes (Figs. 5 and 6).

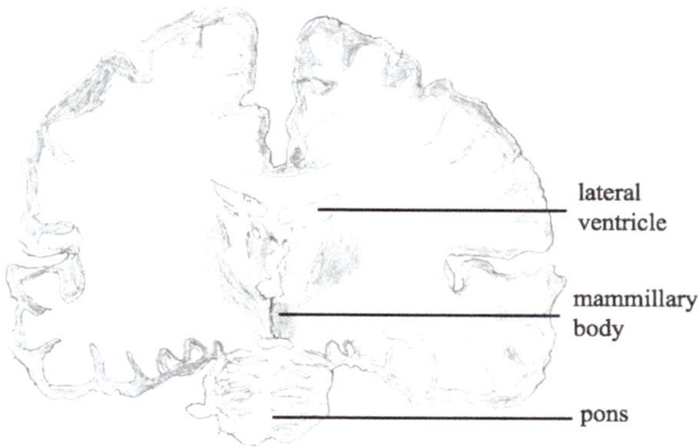

Fig. 5 Schematic cartoon at the level of the mammillary bodies

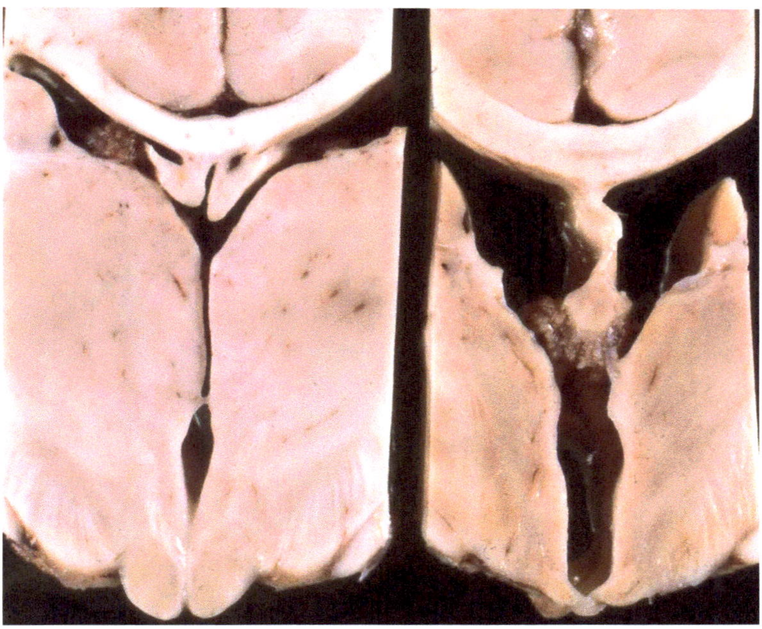

Fig. 6 Brain slices at the level of the mammillary bodies. Normal on *left*; Wernicke's disease on *right*. Figure courtesy of Dr. Roland Auer (see acknowledgments)

In the brainstem, there was gliosis in the periaqueductal areas in over one half of cases, and similar changes in nuclear areas such as the occulomotor nuclei, inferior olives, ventricular nuclei, etc. In the cerebellum, degenerative changes were present in one-third of the cases. Changes were located in the Purkinje cell layer, wherein Purkinje cells were absent. There was an increase in glial cells in the molecular layer (Bergmann) of the cerebellum, and marked gliosis of white matter. These degenerative changes were most pronounced in the vermis.

The cerebral cortex showed no specific alterations. There were changes in a small number of cases which included a diffuse proliferation of large cells called Alzheimer's glia. Several other cases had diffuse neuronyl involvement including, swelling, central chromatolysis, and nissl bodies concentrated at the periphery of the cells. The spinal cord showed pathology in five of 12 cases examined. These cases had diminished numbers of anterior horn cells. Those present had axonyl changes, which were characteristic of peripheral neuritis.

This study of clinical and pathological changes in the Wernicke–Korsakoff syndrome represents the second and largest such study. The pathological changes consisting of lesions localized in the periventricular gray matter, mammillary bodies, and medial regions of the thalamus (dorsomedial nuclei) are nearly identical to the findings reported much earlier in the only other comprehensive study (Gamper, 1928). Intracerebral hemorrhage was rarely found in either study (Figs. 7, 8, 9, 10, and 11).

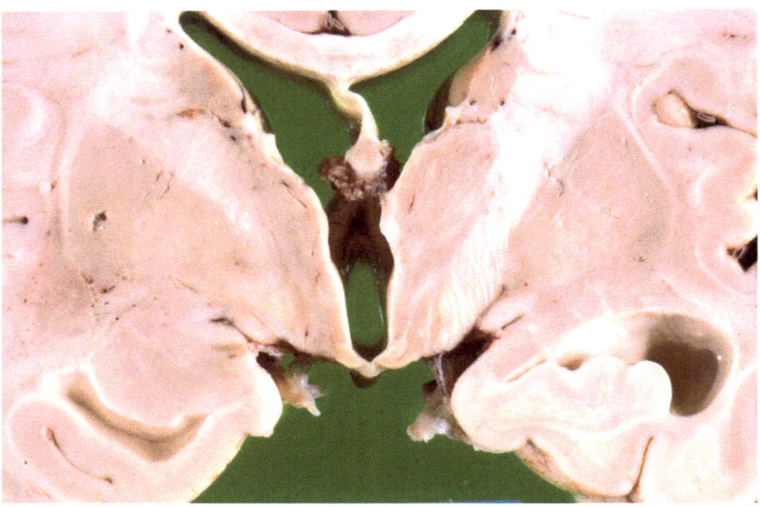

Fig. 7 Close-up of Wernicke's disease mammillary bodies in Fig. 6. Figure courtesy of Dr. Roland Auer

The authors conclude that the lesions seen in their 70 cases of Korsakoff's psychosis correspond exactly with lesions described in Wernicke's disease. This suggests the two disorders are identical, except that one is acute, the other chronic, and that they represent diseases that result largely from thiamine deficiency. Since

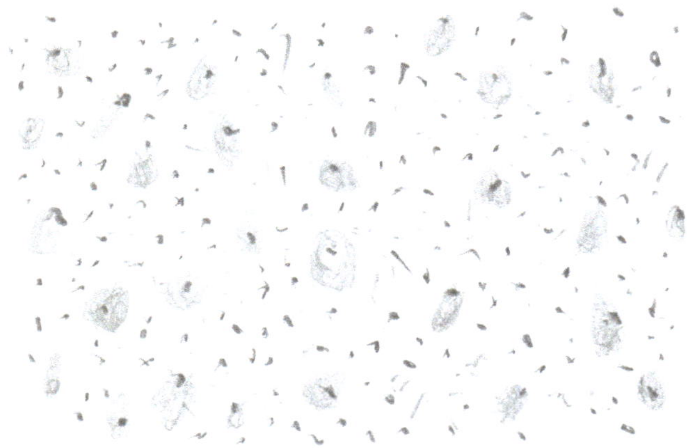

Fig. 8 Drawing of cresyl violet stain section through the mediodorsal thalamic nucleus: control

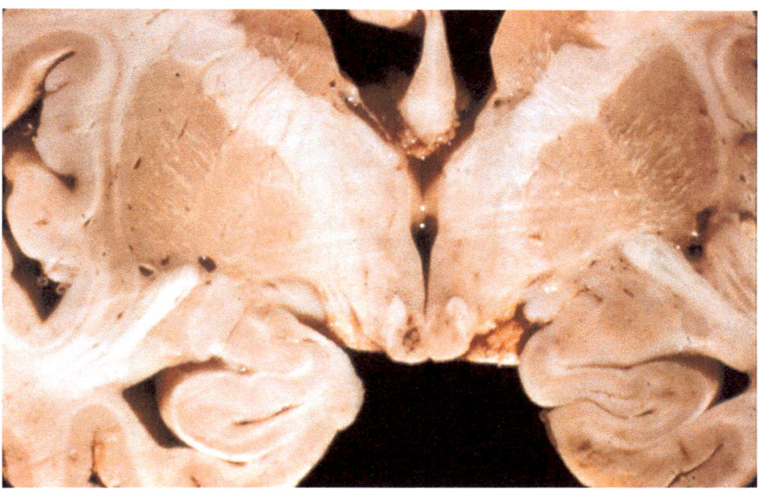

Fig. 9 Wernicke's brain slice at the level of the mammillary bodies. Note lesion bilaterally symmetrical. Figure courtesy of Dr. Michael Norenberg (see acknowledgments)

autopsy data are increasingly difficult to obtain, recent neuropathological data can, in some measure, be obtained from MRI studies, and these studies do show a good correlation with autopsy gathered data (Liu et al., 2006). These MRI studies show bilaterally symmetric lesions in paraventricular regions, hypothalamus and thalamus, mammillary bodies, and periaqueductal areas of the midbrain (Doherty et al., 2002). It should be remembered that neuroimaging studies may not always be

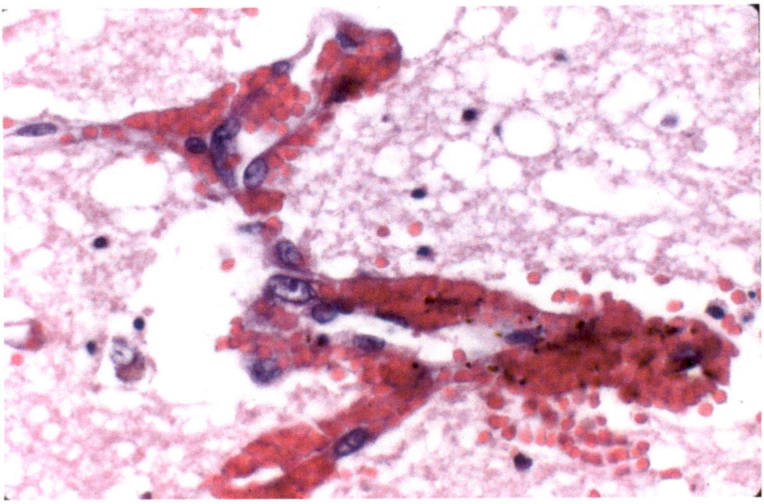

Fig. 10 Wernicke's photomicrograph showing spongy degeneration and presence of blood. Figure coutersy of Dr. Michael Norenberg

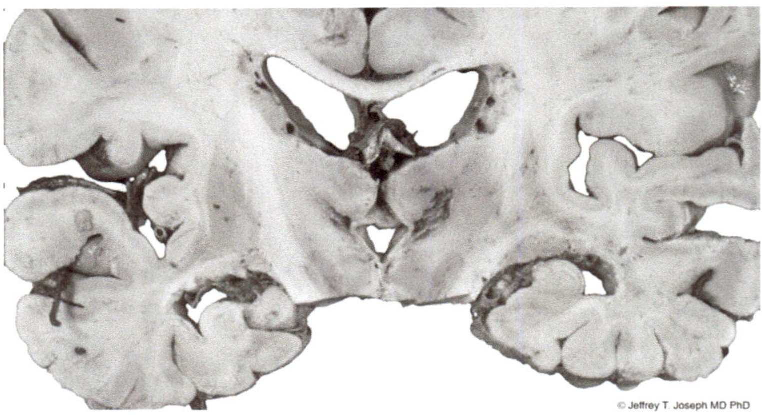

Fig. 11 Brain slice showing mediodorsal thalamus in Wernicke's disease. Figure courtesy of Dr. Jeffrey Joseph (see acknowledgments)

definitive and that they may miss some lesions because of resolution considerations, and because early pathology is biochemical in nature (Celik and Kaya, 2004).

The hemorrhagic component of Wernicke's disease has been examined (Rosenblum and Feigin, 1965). These authors looked at 43 cases they had seen, and two of these had large intracerebral hemorrhages in the vicinity of the third and fourth ventricles. These two cases were classical Wernicke's disease in terms of clinical presentation and other neuropathological findings. The exception was that large (6–8 mm) hemorrhages were located on the side walls of the third ventricle and in the floor of the fourth ventricle. Smaller petechial hemorrhages were located in close proximity. Microscopically, there was vascular proliferation and astrocytosis. The hemorrhagic areas appeared as large lakes of red blood cells. In both cases, the mammillary bodies were affected in the classic way – degeneration including rarefaction, and infiltration of macrophages.

Clinical findings in this paper are of interest in that they showed 90% of patients were chronic alcoholics, and all showed memory impairment and confusion. One-third or more showed ataxia, peripheral neuritis, and findings suggestive of chronic Wernicke's disease. In one case there was a clinical picture which was mixed, with symptoms of both Wernicke's and Korsakoff's diseases. Another case was one of an acute attack of Wernicke's disease which left the patient with loss of memory, disorientation, and confabulation, which lasted until his death, and was reminiscent of Korsakoff's psychosis. These cases show the "overlap" of clinical symptoms, and of pathological lesions. Of the 70 cases, the authors state that 31% showed, in addition to Korsakoff's psychosis, symptoms of a complete or partial Wernicke's syndrome. The pathogenesis of the mental symptoms is not well understood; however, the lesions in the hypothalamus and thalamus have been implicated (Hess and Akert, 1955) as involved. While intellectual functioning occurs in the cerebral cortex, these activities may be integrated in the limbic system, which includes the thalamus and hypothalamus. This could explain many symptoms described in the Wernicke–Korsakoff syndrome.

In the large study of Wernicke–Korsakoff syndrome described earlier (Victor et al., 1989), mental confusion was the most common presenting symptom in two-thirds of the cases (108 of 163). Ataxia was present in over 50% of cases and ocular symptoms were present in 40% of cases. Other symptoms included memory loss, polyneuropathy, exhaustion, etc.

Findings on physical exam in 245 patients with Wernicke–Korsakoff syndrome included tachycardia (51%), fatty hepatosis (49%), redness of tongue (36%), skin and mucous membrane disorders (29%), etc. Hypothermia was observed in only two of the patients in this series; however, a higher incidence of hypothermia has been noted in other reports (Harper, 1979) (Wallis et al., 1978).

Findings related to brain lesions consisted of confusion, defective memory, ocular abnormalities, and ataxia. In this series, confusion could be accurately assessed in 229 patients. Ten percent of these 229 exhibited no alteration in mentation and lacked any confusion. The remainder of patients showed significant changes in consciousness and in mentation.

In this study, 16% presented with symptoms of delerium tremons. These symptoms were usually mild, but clearly related to alcohol withdrawal. Fourteen percent of patients showed little or no confusion or agitation when first examined. The patients did have severe memory and learning disorders characteristic of Korsakoff's psychosis. The majority of patients (56%) had mental confabulation and disorientation that the authors termed "global confusional state." The major characteristics of this were apathy, drowsiness, memory loss, and disorientation. These patients were inert and indifferent to people, questions, and their surroundings. Other symptoms included memory disorder, alcohol abstinence state, stupor, and coma.

Ocular symptoms occurred in 96% of patients. These symptoms consisted of nystagmus, lateral rectus palsy, and paralysis of conjugate gaze. Each patient had more or less frequency of these symptoms; only 27% had all three symptoms present. Other ocular symptoms were uncommon. A sluggish reaction of the pupils to light occurred in 19%, yet it was relatively mild.

Nystagmus as a presenting symptom was noted in 85% of patients. The most common type was horizontal nystagmus to both sides. This was seen in 97% of patients with nystagmus. One half of patients with horizontal nystagmus showed vertical nystagmus. Treatment of nystagmus with thiamine cleared nystagmus in as little as 3 days. In 54% of cases, a lateral rectus palsy was present on first examination. The majority of these patients had incomplete lateral rectus palsy. Weakness of conjugate gaze was present in 44% of patients examined. Eighty-six percent of those with gaze palsies presented with a horizontal plane gaze palsy.

Of the patients examined initially, only 13% showed no alteration in ability to walk or stand. At the other end of the spectrum, 21% of patients could neither walk nor stand unaided. These patients took uncertain lurching type steps and wanted only to return to bed. Mild ataxia consisted of only being noticeable when the patient was asked to perform tasks such as heel to toe walking. A characteristic tremor of the lower extremity was noted, as was an incoordination of the upper extremity. A few patients showed a combination of ataxic gait and ataxia of both upper and lower extremities (Figs. 12, 13, and 14).

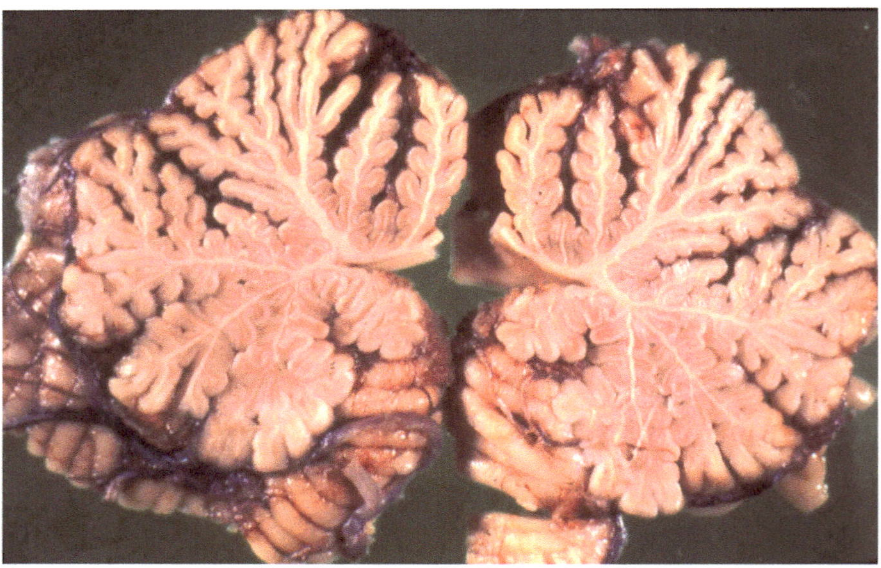

Fig. 12 Cerebellum in Wernicke's disease. Figure courtesy of Dr. Roland Auer

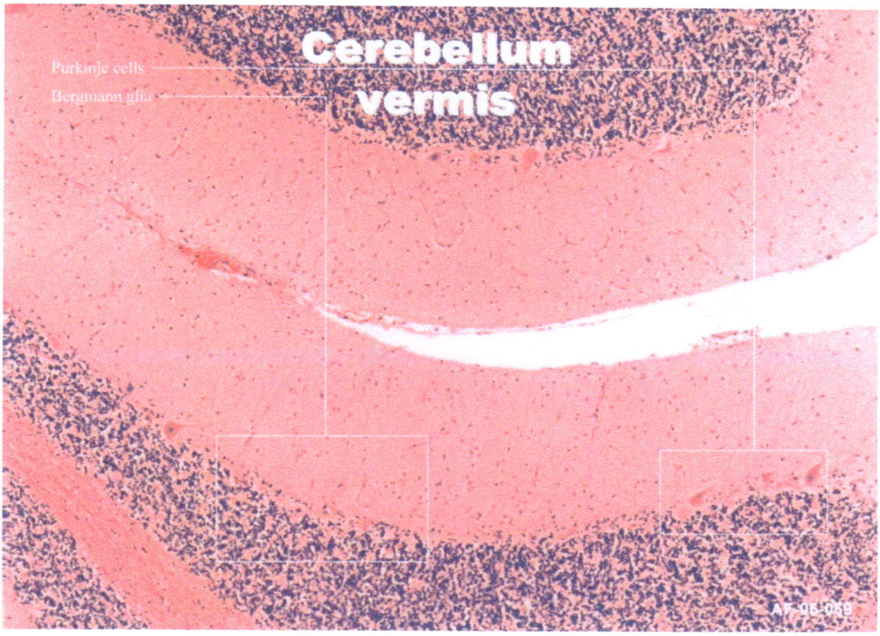

Fig. 13 Photomicrograph of normal control cerebellar layers showing Purkinje cells and Bergmann glia. Figure courtesy of Dr. Roland Auer

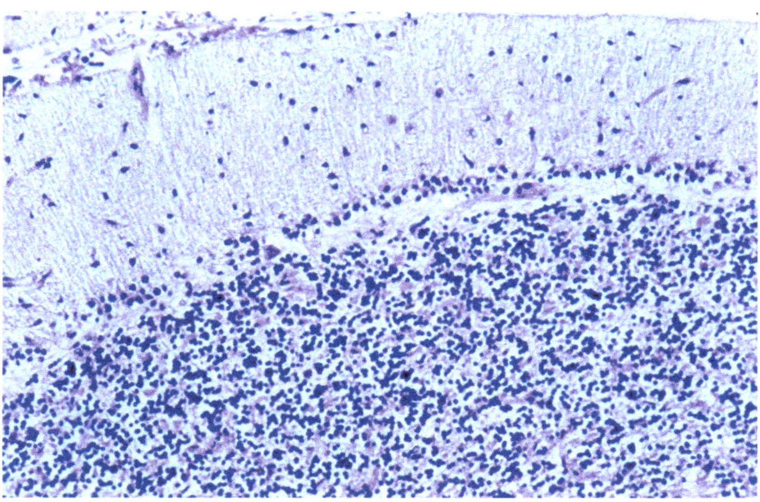

Fig. 14 Photomicrograph of cerebellar layers – note the absence of Purkinje cells. Figure courtesy of Dr. Roland Auer

In 82% of patients, polyneuropathy was present. The legs were affected in most cases, but in some, both upper and lower extremities were affected. In symptomatic cases, complaints consisted of weakness, pain, and paresthesias. Symptoms were usually noted first in distal portions of the extremities and then progressed proximally. In some cases, the polyneuropathy was notable only on physical examination. The exam found tender leg muscles, decreased ankle jerks, and in some instances, loss or decrease in knee-jerk reaction. In a few cases, the upper extremities were affected as well. These patients had loss of tendon reflexes and loss of motor strength. Muscle weakness often took the form of foot and/or hand drop. Not only were distal muscles affected but proximal muscles were also weakened. Total paralysis was rarely observed. Tenderness to pressure was almost always noted, usually in the calves and feet. Muscle weakness in the legs and feet were almost always associated with diminished knee and/or ankle jerks. Edema and thinness of skin were common findings over the lower legs and feet in severe cases of polyneuropathy in the lower extremity.

In terms of cranial neuropathy, except for occulomotor and vestibular function, other cranial nerve involvement does not occur. Bulbar palsy also does not occur in the Wernicke–Korsakoff syndrome. A few cases of dysphonia and dysphagia have been described (Novak and Victor, 1974). These were cases where there was degeneration of the vagus nerve, and in which the patients also had a severe nutritional deficiency component in their history. For a recent review, see Kashi, M. et al., (2009).

Abnormal laboratory findings were also noted at hospital admission. Abnormal liver function tests were present in 43% of patients, and anemia was found in 39%. Other laboratory chemistry abnormalities included leukocytosis, elevated BUN, hyperglycemia, and hypovalemia. EEG results were available in a low number of cases, and in half, there was a diffuse decrease in the frequency of brain waves. This change was only mild to moderate. Remarkable was the finding that in spite of obvious neurological involvement, half of those tested had normal EEGs. Some of these "normal EEG" patients had extensive diencephalic lesions, and reduced cerebral blood flow, with reduced O_2 and glucose consumption. Examination of CSF revealed elevated protein levels in 23% of patients. Other than protein levels, all CSF values were within normal values.

As stated above, recovery following thiamine treatment produced a rapid recovery of function of some deficits. In the case of abducens nerve palsy, recovery of function was noted as soon as 6 hours after thiamine administration. In another group, recovery was noted in less than a day, and in all cases, improvement was seen in less than 4 days. Complete recovery had occurred in most patients in less than 7 days. Vertical gaze palsy showed improvement within 1–24 hours in half of the patients. Other ocular symptoms showed similar modes of recovery.

Ataxias were somewhat slower to show recovery. In most cases of ataxia that could be observed properly, recovery was noted in 2–6 days. In cases where recovery could be determined, complete recovery occurred in less than 1 month in almost half of the cases. Most recovered cases occurred in less than 2 months, but only 38% showed complete recovery. No significant improvement in ataxia occurred in

27% of patients. The remaining group made an incomplete recovery, plateauing at a stage of lessened ataxia. These improvements were usually completed by the end of the second month.

Thiamine administration had a positive effect on the recovery from the confusion and confabulation state in basically all patients who were properly examined. The onset of improvement in mental symptoms occurred by 17 days after thiamine administration. In some few cases, complete recovery from the confused state occurred in only 6 days; others took up to 2 months. In patients with amnesia in Korsakoff's psychosis, about 25% in each of four groups showed no recovery, slight recovery, significant recovery, or complete recovery. The mechanism by which some recovered completely, and others not at all, is unclear, but may be related to the length of time the patient had been sick. It should be noted that, in this study, 10% of patients had no mental symptoms. Also, in about 10% stupor/coma was present, making further assessment difficult.

Well-controlled treatment studies have been conducted in some measure because of the large number of available Wernicke-Korsakoff patients. In one such study (Phillips et al., 1952), 51 patients were studied, and 29 served as well-fed, well-nourished control subjects. Experimental subjects received no food or nourishment until their clinical status had been carefully assessed. The experimental patients were divided into four groups: (1) glucose plus salt water, (2) an unfortified rice diet, (3) a vitamin-deficient diet, and (4) a group who received the above three plus whiskey.

Results showed that in the first group, there was no improvement in the clinical picture until thiamine was added to the diet. The addition of a full supplement of vitamins excluding thiamine had no effect. In the second group (unfortified rice diet) symptoms remained unchanged until thiamine alone was added. The third group also had no improvement. A subgroup of these patients was also given whiskey, and after thiamine administration, they also improved. The memory loss was slow to recover compared to recovery from nystagmus, but improvement of memory was evident by 8 weeks in most patients. This is in keeping with the slower recovery of mentation seen in nearly all patients with Wernicke–Korsakoff syndrome.

These studies (Phillips et al., 1952) establish beyond doubt that the symptoms of Wernicke–Korsakoff syndrome can be directly related to thiamine deficiency. Similar studies by others (Victor and Adams, 1961) confirm these findings. The studies emphasize the importance of instituting thiamine therapy as soon as possible in order to decrease the biochemical lesion, and prevent as much structural damage as is possible. In some cases in which unsupplemented glucose is given, without thiamine, Wernicke's symptoms may be precipitated in a very short time. Patients in whom thiamine is greatly reduced may be pushed into overt disease by a high-carbohydrate diet, thereby depleting whatever residual amount of thiamine was present.

In Korsakoff's psychosis, symptoms are slower to recover and may be more apparent when Wernicke's symptoms clear. Weeks to months may elapse before Korsakoff's psychosis symptoms begin to improve, but eventually most patients with Korsakoff's do have some improvement. As many as 21% will make a complete recovery, and a higher rate has been recorded in other studies (Rosenbaum and Merritt, 1939).

In the early stages of the Wernicke-Korsakoff disorder, treatment with thiamine can show almost immediate benefits, emphasizing the biochemical lesion characteristic of this disease. The actual dose, and length of treatment of patients, however, is not established (Morcos, 2004). The usual dose is of an initial load of 100 mg IV, followed by a maintenance dose of between 50–100 mg by mouth until the patient is eating well (Lacasse and Lum, 2004).

In a paper (Grunnet, 1969), changing patterns of incidence, distribution, and neuropathology in Wernicke's patients were examined. Results showed that the time of death seemed to have lengthened over a period of 25 years, probably due to improved health care. There appeared to be an increase in the numbers of cases, possibly due to increased awareness and availability of health care. Finally, the authors describe a somewhat more chronic form of Wernicke's disease in which the lesions were more widespread, and to some degree, the lesions were more at a microscopic level than at a gross level.

Another study examining the incidence of Wernicke's disease in Australia in 131 cases has been published (Harper, 1983). This study showed that the Australia incidence was (2.8%) higher than any other country. The exact reasons for this are unclear; however, the author may have been more efficient in looking at autopsy material. He points out that many cases are missed at autopsy and that clinical features may not always be obvious. In this study, both the gross and histological changes were quite similar to those described previously (Victor et al., 1989).

In a neurochemical study, (McEntee et al., 1987) the authors had noted that the lesions of Korsakoff's psychosis corresponded with the location of monoamine containing neurons. The lesions and monoamine containing neurons are both located in periventricular and periaqueductal regions of the brain. A total of 25 Korsakoff's patients were studied. CSF was obtained from these patients and controls, and 3-methyl-4-hydroxyl phenylglycol, homovanillic acid, and 5-hydroxyindolacetic acid were measured. Psychometric analysis just prior to obtaining CSF showed a normal intelligence, but a significantly impaired memory function. The CSF levels of the three compounds measured were statistically lower than that of controls. These compounds are a measure of cerebral noradrenergic activity. The compounds are also altered in disorders such as Parkinson's disease and Alzheimer's disease. The possibility exists that these other diseases may have been present in some patients, contributing to the results. However, this study shows a possible link between damage to noradrenergic neurons and the memory losses seen in Korsakoff's psychosis.

In a second study (McEntee et al., 1987), the CSF concentrations of norepinephrine, dopamine, and serotonin were examined in a group of Korsakoff's patients. Measurements were made and correlated with the learning of two motor tasks. Results showed that improvement in learning correlated with CSF levels of dopamine metabolites. These studies indicate that dopamine might be involved in learning motor tasks and that a depletion of CSF dopamine metabolites could hinder learning motor tasks in patients with Korsakoff's psychosis.

In a retrospective study of clinical signs (Harper et al., 1986), careful analysis of signs initially noted in the Wernicke–Korsakoff syndrome are discussed. These authors note that in their study, if the criteria previously described were strictly adhered to, the diagnosis would have been missed.

The authors state that mild cases of Wernicke's disease surely exist and that a less rigid criteria should be applied. Attention should also be placed on suspicion for the diagnosis based on history. It is also noted that many mild episodes may, over time, produce Wernicke's disease in a chronic form with a somewhat different presentation of symptoms. The problem is that if a rigorous standard is applied, then cases of the Wernicke–Korsakoff syndrome may be missed, and treatment delayed, or not applied. If more relaxed criteria are utilized, and suspicious cases identified, then adequate thiamine therapy can be administered. The proof of this awaits a neuropathologic study in which pathologically confirmed cases are identified who had minor symptoms and therefore were not identified.

Wernicke's disease can occur in non-alcoholic patients. This has been described (Jagadha et al., 1987) in a group of five patients receiving dialysis. In all five, there was no indication of alcohol abuse, but each had end stage renal disease. In addition, they were all receiving IV therapy and were very sick. They had repeated dialysis, either peritoneal dialysis or hemodialysis. Infection was another feature, which could compromise thiamine metabolism.

Upon death, neuropathological studies were performed. Brain examination revealed lesions characteristic of Wernicke's disease. These lesions were present in the inferior olivary nuclei, hypothalamus, thalamus, and mammillary bodies. The lesions were bilaterally symmetrical. Microscopic examination revealed nerve cell death, axonyl swelling, gliosis, myelin loss, brainstem vascular proliferation, and hemorrhages. In one case, the hemorrhages were confluent and prominent. Some of the microscopic findings were in areas which showed no gross abnormalities, emphasizing the need for a thorough pathological exam. It should be emphasized that many factors can modify thiamine metabolism and produce Wernicke's disease in very ill patients, such as those undergoing dialysis (Figs. 15 and 16).

The actual levels of activity of key thiamine enzymes have been examined in the brains of patients with Wernicke–Korsakoff syndrome. In this study (Butterworth et al., 1993), control, alcoholics without Wernicke's, and Wernicke-Korsakoff patients' cerebellar vermis were studied. At autopsy, vermis samples were removed and stored at –80°C until analysis. Neuropathological analysis showed that tissue from nine patients that were alcoholic/non-Wernicke's did not show evidence of Wernicke's disease. The two patients with Wernicke–Korsakoff syndrome had small mammillary bodies. Microscopic examination showed the usual spongy characteristic capillary proliferation and gliosis. The gliosis was present in the periventricular region of the third and fourth ventricles.

Results from biochemical analysis of the three key thiamine-requiring enzymes in the cerebellar vermis showed decreases in all three enzymes in the Wernicke-Korsakoff patients. The greatest drop was in the activity of alpha-ketoglutarate dehydrogenase. The non-thiamine requiring enzyme glutamate dehydrogenase was unaltered, serving as an internal control for fixation artifact. These data suggest that

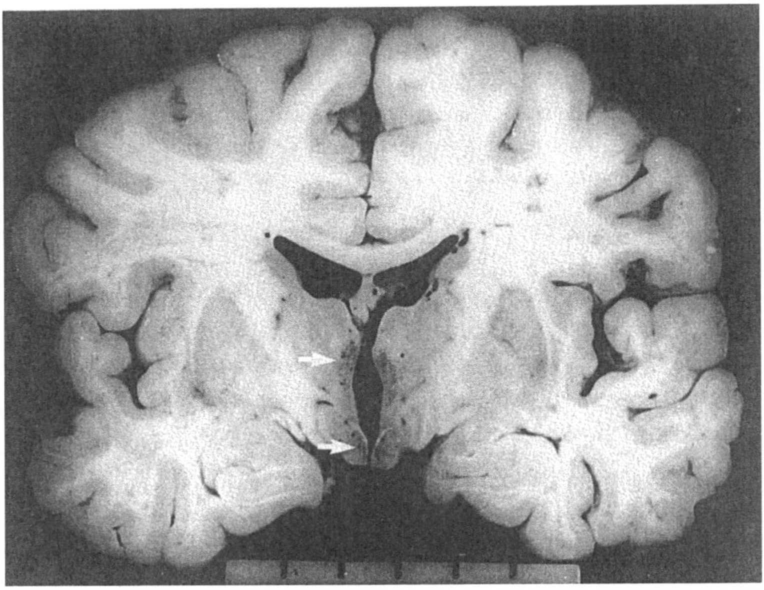

Fig. 15 Brain slice showing lesioned mammillary bodies and mediodorsal thalamic nucleus. Reproduced from Krill (1995) with permission from Springer

significant decreases in thiamine-requiring enzymes may be essential in the disease process, and their alterations may in turn help explain symptoms in these patients.

In another study examining treatment modalities (Ambrose et al., 2001), thiamine was administered to various groups of alcoholic-dependent patients. These patients were undergoing detoxification and were administered thiamine at different dose levels. This was done without heavy dependence on the classic triad of symptoms: ataxia, ocular disturbances, and confusion. These strict symptoms, when used, may exclude some patients with Wernicke's disease.

The investigators examined patient's performance on a mini-mental state examination. Of the various treatment groups, the group receiving the highest dose of thiamine (200 mg) performed highest. Other factors such as education, years of drinking, etc., were not significant contributors to exam performance. These data reinforce the concept that patients with mild, less symptomic forms of Wernicke's disease benefit from thiamine therapy. These patients should be actively looked for, recognized, and treated.

A recent study was undertaken to evaluate MRI findings in light of signs and symptoms in a group of Wernicke's patients (Zuccoli, G. et al., 2007). This was a retrospective study of 14 female and 12 male patients. Inclusion was based on a clinical diagnosis of Wernicke's disease, and of significant improvement after initiation of thiamine treatment. All 26 patients had MRI performed, which included long repetition time and short echo time spin-echo sequences, and contrast-enhanced short repetition images in multiple planes.

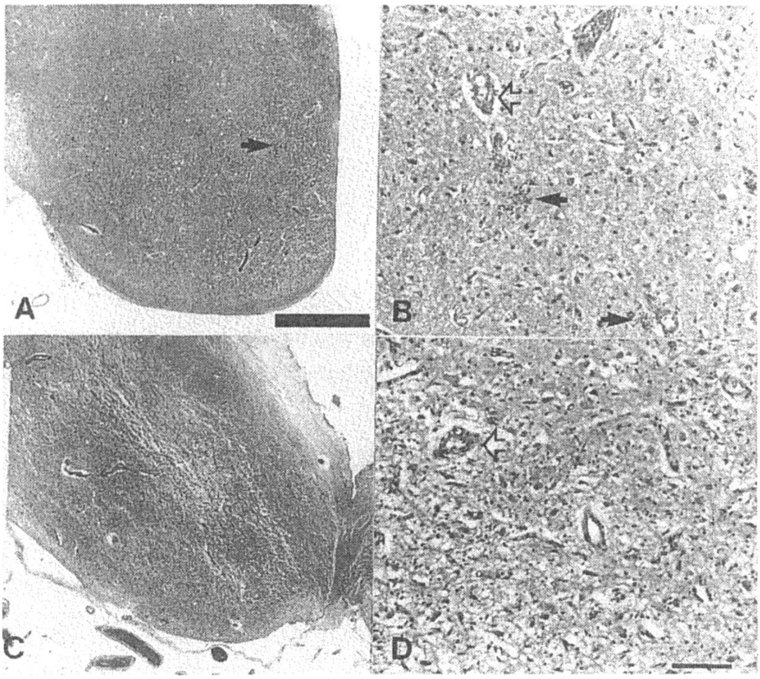

Fig. 16 Photomicrographs of lesioned areas shown in Fig. 14: mammillary bodies and mediodorsal thalamic nuclei. Reproduced from Krill (1995) with permission from Springer

Results showed that one half of the patients had alcohol abuse, whereas one half did not. In the non-alcohol abuse group, Wernicke's was associated with GI tumors. Other causes included voluntary starvation, anorexia nervosa, and poverty. The most prevalent symptoms (81%) were changes in the level of consciousness. This ran a range of mild disorientation to coma. Twenty patients (77%) had ocular symptoms, while 14 (50%) patients were ataxic. Only the ataxic patients had a statistical correlation with the alcoholic group.

Results from MRI studies showed that 23 patients had high signal intensity changes on long repetition time spin-echo sequences, and eight patients had low signal intensity changes on short repetition time spin-echo images. Twenty-two patients had symmetric lesions in the medial thalamic nuclei and periventricular region of the third ventricle. Fifteen patients had lesions in the mammillary bodies. Only one patient showed lesions in the cerebellum.

The authors speculate that although the clinical triad of Wernicke's disease is frequently met (nystagmus, ataxia, and confusion), incomplete neurological presentation also occurs. The anatomic regions most frequently affected in this retrospective study were the medial thalami and the periventricullar region of the third ventricle. The authors suggest that these areas are where the maintenance of osmotic gradients is closely related to thiamine levels, and so decreased thiamine might be expected to produce altered physiology, and onset of lesions. Alteration of structure in the mammillary bodies correlated with the alcohol-abuse group of patients (Fig. 17).

Classical Clinical Findings in Wernicke's Disease
Nystagmus
Confusion
Ataxia

Fig. 17

Wernicke's disease can occur in infants and has recently been described in six infants who were fed a soy formula which had no thiamine content (Kornreich, L. et al., 2005). These infants all were admitted at about the same time. Blood lactate was increased, but there was no evidence of infectious agents. MRIs were similar, and soon a connection between the MRI details and a certain soy formula was made. After analysis of the formula showed no thiamine, thiamine treatment was initiated and improvement was noted.

MRI examination showed T2-weighted areas of hyperintensity in the mammillary bodies, periaqueductal gray, thalami, basal ganglia, and brain stem. All lesions were bilateral and symmetric. These results emphasize that the Wernicke's diagnosis showed lesions in the same general areas as are seen in adult patients, especially lesions in the mammillary bodies. MR images demonstrated a characteristic lactate peak.

Central Pontine Myelinolysis, Alcoholic Cerebellar Degeneration, and Marchiafava–Bignami Disease

Central Pontine Myelinolysis

Central pontine myelinolysis (CPM) is a distinct disorder with characteristic CNS neuropathology and associated with thiamine deficiency. This is not a common disorder and is often associated with some other severe illness independent of alcoholism.

In one series of cases described in detail (Adams, R. et al., 1959), a patient was admitted with a long history of chronic alcoholism. Upon admission, the patient was seriously ill, had been drinking heavily, and was confabulating. He had chest pain and a cough. Culture revealed *Diplococcus pneumoniae* type 5. He was given penicillin, saline, carbohydrate, and thiamine. For 3 days, he improved and then became tremulous and confused. He again improved and then by day 9, he became unable to swallow or move his extremities. By day 11 he was flaccid and did not react to painful stimuli. On day 12, he was responsive to stimuli. He showed no nystagmus and could open and close his mouth, but could do little else. There were no deep reflexes at the wrists, knees, or ankles. By day 14, there was improvement in reflexes, but no other improvement. A chest X-ray showed diffuse mottled density in both lungs. Respiration became shallow and gasping, and death from respiratory arrest occurred on the 22nd hospital day.

Neuropathological features included convolutional atrophy and thin leptomeninges. Major cerebral blood vessels were normal. The external surface of the pons was gray and soft. Cortical parietal lobes were swollen and had congested vessels.

The pons showed a striking symmetric punctate area upon myelin staining. This lesion extended through the middle and upper pons, affecting almost the entire basis pontis. Superiorly, the lesion extended into the tegmentum just to and into the medial lemnesci. The lesion edges were clearly demarcated, but were irregular. Bundles of myelinated fibers which entered the lesion became thinned and pale. Scattered in the lesion were remnants of macrophages containing hemotoxyphilic granules. The largest macrophages were at the periphery of the lesion. Nerve cells and axis cylinders were largely unaffected. No oligodendrocytes could be seen in areas of demyelination. In these areas, macrophages were fat laden. Some astrocytes were

D.W. McCandless, *Thiamine Deficiency and Associated Clinical Disorders*, Contemporary Clinical Neuroscience, DOI 10.1007/978-1-60761-311-4_6,

present. Within the lesion, blood vessels seemed engorged. No other lesions or cells similar to those in the lesion were found elsewhere in the brain. A few other lesions thought to be inconsequential included those found in the spinal cord columns of Goll, in the medulla in the floor of the fourth ventricle, and in the dorsal motor nuclei of the vagus nerve. There were small hemorrhages in one mammillary body (Figs. 1 and 2).

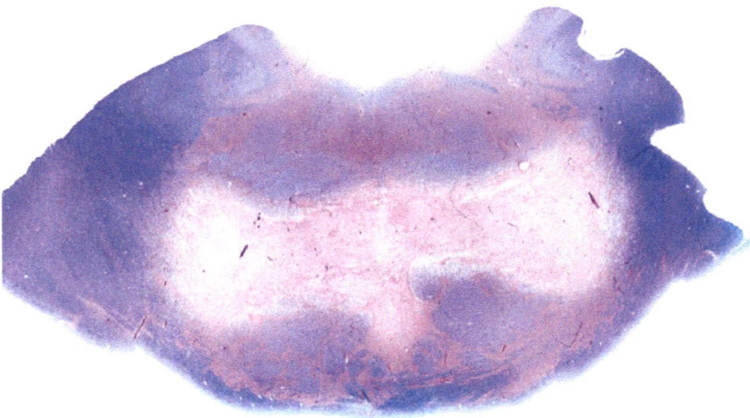

Fig. 1 Photomicrograph of central pontine myelinosos showing massive demyelination. Figure coutersy of Dr. Michael Norenberg

In another case, described by the above author, a malnourished alcoholic woman was admitted to the hospital dehydrated and anemic. Over time she developed flaccid quadriplegia and paralysis of bulbar and respiratory musculature. The neurological symptoms continued to get worse, and she expired of respiratory paralysis.

Neuropathologic findings showed a normal external appearance. The basal portion of the pons was soft and gray. Microscopically, lesions were present in the rostral portion of the basal pontis. The lesion extended into the tegmentum and involved the medial parts of the medial lemnesci. The lesion was sharply demarcated, but irregular. Again, myelinated fiber bundles approaching the lesion were normal, then quickly thinned out, and became pale. Nerve cells and axis cylinders were mostly spared in the lesion. Neuron numbers were slightly decreased. Some neurons were smaller and/or pale. Macrophages were engorged with fat. Oligodendrocytes were not seen, while a few large astrocytes were noticed. Minor changes were seen in other areas, including the anterior horns of the spinal cord, inferior olivary nucleus, and nucleus ambiguous.

A third case was a malnourished alcoholic 43-year-old man who developed a flaccid weakness of both legs. He also had advanced pulmonary tuberculosis. On admission, he had an absence of knee and ankle jerks. He was anorexic, and an X-ray showed a pneumonic density in the right lung. The patient's temperature and pulse remained high, and he expired on the 11th hospital day.

Fig. 2 Photograph of central pontine myelinolysis showing demyelination, but with Sparing of the cortical spinal tracts. Figure courtesy of Dr. Michael Norenberg

Neuropathological findings were normal regarding superficial appearance, and the meninges were also normal. The basal area of the pons was gray. The lesion was largely restricted to the central area of the pons, but was smaller in size than the first two cases. There was a sharply demarcated focus of pallor in the basis pontis. The lesion had many macrophages loaded with granules. The lesion was only about 4 mm in diameter, was sharply demarcated, and within the lesion no normal myelin sheaths could be seen. Axis cylinders were decreased in numbers. Nerve cells were normal in appearance and numbers. A few normal astrocytes were seen, and also some enlarged astrocytes were noted. Engorged capillaries and venules were seen in the lesion.

A few other alterations were seen in other areas, such as "loosening" of tissue matrix and an increase in number and size of astrocytes. Neurons and blood vessels were normal in the mammillary bodies. Other areas with changes included the walls of the third ventricle, medial nucleus of the thalamus, third nerve nuclei, red nucleus, and in the periaqueductal gray matter. The lesions in this case were considered characteristic of Wernicke's disease. This case also had prominent astrocytes in the cerebral cortex and dentate nucleus.

The authors commented on the remarkable localization and symmetric distribution of lesions in these three cases. In the first two cases, the majority of the basis pontis was involved, whereas in case 3 a smaller portion of the basis pontis was involved. Each case was one of demyelinative nature. Within the area of the lesion, almost all medulated sheaths were involved. Thus, deyelination was present in all tracts running through the lesion. In spite of an advanced degree of myelin destruction, axis cylinders, neurons, and blood vessels were preserved.

Most neurons were preserved, although some were swollen, and in some areas, neurons seemed to be reduced in numbers. These neural changes were consistent with those seen in areas with interruption of axis cylinders (axonyl reaction). Blood vessels were patent in the lesions, but small vessels may have had increased cellularity in the walls. There was a lack of oligodendrocytes, only being found in the very margins of the lesions. There was an increase in astrocytes, characteristic of myelin degeneration type lesions. Macrophages were seen containing large refractive non-hematoxyphilic particles.

The authors state that the lesion must have originated near the center of the pons based on the appearance of cavitation at the center. Also, the most recent areas of neuropathology were at the periphery. These lesion features consisted of large pieces of degenerating myelin, and phagocytes. This recent lesion activity was around the complete periphery. The authors speculated that the central (oldest) portion of the lesion was about 6–8 weeks old. At least some of the advancing lesion correlated with symptoms.

The authors made some comparisons between symptoms and lesions. All three cases had a history of severe alcoholism lasting many years. Many symptoms could be attributed to the lesions in the pons. Thus, flaccid quadriplegia, difficulty in speaking and swallowing, decreased response to painful stimuli, and loss of corneal responses could all be caused by the lesion. Cortico-bulbar tracts pass through the pons, and their loss would affect speech and swallowing. Loss of corneal reflexes would be attributed to trigeminal nerve loss.

The authors reject the idea that this is merely an unusual example of Wernicke's disease because of the characteristics of the lesion. In all cases of Wernicke's, the mammillary bodies had been affected, as well as the walls of the third ventricle, peri-aquaductal region, and the floor of the forth ventricle in most cases. Also, histologically, the lesions were more generalized and not limited to demyelination.

The lesions of CPM also differ from those of Marchiafava–Bignami disease in that the location of lesions is most likely seen in the corpus callosum. Microscopically, the lesions of Marchiafava–Bignami disease are those of demyelination and some effect on axis cylinders, and the lesion can progress to cavitation. One other feature of Marchiafava–Bignami disease is the significant role of nutrition, in that several cases have been confirmed in patients with no history of alcoholism.

The neuropathological features of symmetry and constancy of location seem to indicate a biochemical base for etiology. This is meant to say that some toxin or a deficiency of some vital nutrient essential to the well-being of the brain is at the seat of the disorder. Examples of these include thiamine deficiency, as well as cyanide poisoning, anoxia, Wilson's disease, hepatic encephalopathy, and hypoglycemia.

Several years later after the first description of CPM, the rarity of this disorder was recognized in that only 18 cases had been described in the world, and none were diagnosed before death (Cole, M. et al., 1964). These authors suggested that malnutrition is a critical feature and present two new cases to support that contention.

The first patient had a long term history of alcoholism and irregular dietary habits. Just before hospital admission, the patient had a sore throat and mouth and facial swelling. After admission, her condition remained unchanged when she complained of diplopia. She was found to have weakness of all eye movements. A couple days later, she choked while eating lunch and expired.

Necropsy showed an acute moderate pulmonary congestion. The brain was grossly normal with no evidence of arterial disease. Microscopically, the brain showed rounded and pale Betz cells in the cerebral cortex. The Betz cells seemed normal in number. The hippocampus, hypothalamus, and mammillary bodies and midbrain were normal in appearance. However, centrally located in the basalis pontis was a sharply demarcated bilateral lesion with an almost total lack of any myelin staining fibers. Beyond the periphery of the lesion where the tissue was normal, the myelin was intact and stained well. Axis cylinders were present within the lesion, but were poorly stained. Neurons were absent at the center of the lesion, while at the periphery, neurons were swollen and poorly stained. Nuclei in the reticular formation were rounded, swollen, and had central chromatolysis.

The second case was a man admitted with slurring of speech, diplopia, and weakness. The patient had a history of heavy drinking. Several years earlier he had had a pulmonary lobectomy for pulmonary tuberculosis. He had a severe drinking spree just before hospitalization; he then refused to eat. When admitted, he was confused and dysarthric. Diplococcus pneumonia was cultured from sputum. He was fed IV and given multivitamins including B1. He continued to deteriorate and expired 1 month after admission.

Upon autopsy, the brain was normal externally. Histology showed normal cerebral cortex, cerebellum, and basal ganglia. The mammillary bodies had a decreased number of neurons along with an increase in microglia and capillary prominence.

Centrally in the basalis pontis was a sharply demarcated symmetric lesion in which there was an absence of myelin. As in previous cases, the myelin was normal right up to the lesion's edge. Neurons were normal in appearance in the lesion, but there was a decrease in oligodendrocytes.

Both of these two cases were in chronic alcoholics, but in both cases, there was significant malnutrition. There were neurons in the cerebral cortex and brain stem consistent with a nutritional deficiency called "central neuritis," and seen in brains from pellagra patients. These neuronyl abnormalities have been seen in other cases of severe prolonged malnutrition.

The authors also point out that in 17 cases of CPM presented by Berry and Olszewski (Berry, K. and Olszewski, J., 1963), ten had evidence of malnutrition. Most other patients had evidence of mild malnutrition. Some cases of CPM have been seen in which there is no evidence of malnutrition. The exact nature of the malnutrition seen in these cases is not clear, and there is also some variation in presenting the clinical picture. Symptoms suggestive of brain stem lesions such as nystagmus, quadriplegia, dysarthra, dysphagia, extraocular palsies, and stupor and coma should alert clinicians and facilitate diagnosis.

By 1976, and the paper by Tomlinson, B. et al. (1976), the number of cases published was 120. Some cases described demyelination in other areas besides the pons. Several concurrent diseases were described including Wilson disease, amyloidosis, and liver disease, as well as alcoholism and malnutrition. The paper reviews the literature and presents two more cases of CPM.

The first case was of a 54-year-old female admitted in a confusional state with a several-day history of vomiting, perhaps due taking to phenybutazone. Little other coherent history could be obtained. The patient's conscious level declined; she became unresponsive to deep stimuli and had generalized hypertonicity and exaggerated reflexes. She continued to deteriorate, became stuporous, and died 5 weeks after admission.

Necropsy showed normal brain exterior except for slight granularity over the surface. There was severe demyelination throughout the dorsal pons to the middle cerebellar peduncles. The area of the lesion appeared gray. There was a bilaterally symmetric lesion around 1 mm each in the putamen.

Histologically, the main lesion in the pons showed extensive pontine demyelination. Moderate astrocytic proliferation was present in the lesion, and oligodendrocytes were not seen. The lesions microscopically were well delineated. The thalamus had demyelination, although not as severe as in the pons. Lesions were also located in the caudate nucleus, claustrum, putamen, hypothalamus, and scattered cortical areas. White matter in cerebellar folia was demyelinated and exhibited astrocytic and microglial infiltration.

The second case in this paper was a 57-year-old woman admitted because of drowsiness. She had had a 3-week period of vomiting. She later developed seizures and expired. At autopsy she showed terminal bronchial pneumonia and bilateral suppurative pyelonephritis.

The surface of the brain showed slight convolutional atrophy. Vessels at the base of the brain appeared normal. On section, a central pontine gray area extended from the inferior colliculi to the mid pons. Other sites included the cerebral cortex, basal ganglia, and thalamus.

Microscopically at mid pontine level, there was almost total demyelination. There were neurons present, but they were reduced in number. There were cavities of total tissue destruction with debris and a few macrophages. This lesion ceased abruptly in the lower pons, and the medulla was normal. A few lesion areas were present in the cerebral cortex, but most neurons were preserved. The cerebellum also had a few areas of necrosis, along with marked astrocytic infiltration (Fig. 3).

The authors point out that the lesions in other reported cases are strikingly similar, while the actual size of the lesion varies from case to case. Some central pontine lesions are so small (less than 6 mm) that they could easily be missed on sectioning. There is also a wide variation in the size of the lesion and the clinical symptoms. Some patients had large lesions and minor symptoms, and some patients had small lesions and major symptoms. Many had evidence of other CNS disorders, especially Wernicke's disease. Many cases have also been associated with other chronic illnesses, especially pulmonary. These two cases did not present with other major diseases. These two cases did have lesions outside the pons, as have others.

Central Pontine Myelinolysis—Microscopic Lesion

Source	Convolutional atrophy	Soft pons	Myelin thin and pale	Nerve cells affected	Increased glia	Blood vessels engorged
Adams, Victoe, & Mancall	+	+	+	–	+	+
Cole, Richardson, & Segarre			+	+	+	
Tomlinson, Pierider, & Bradley	+	+	+	+	+	

Fig. 3

There is estimated to be about 10% of cases with more widely distributed lesions. The two patients in this study had strikingly electrolyte disturbances (hyponatremia, hypokalemia, and hypochloremia). These electrolyte disturbances have been noted in most cases of CPM in the terminal phase, suggesting a contributory effect.

Speculative reasons for the highly selective nature of lesions in CPM are discussed. One reason stated is that the venous drainage could be insufficient, such that poor drainage could predispose for disease. Others have suggested that the oligodendrocytes and astrocytes were susceptible to disease and toxic/metabolic problems. This could be a problem with the proximity of nerves (demanding good circulation) and fiber bundles. This is strengthened by the location outside the pons of lesions. These invariably involve brain areas where there is a close association of neurons and fiber bundles such as the claustrum and internal capsule. The authors also say that the close proximity of oligodendrites and neurons could be detrimental in that neurons in stress could "steal" blood supply from oligodendrites and myelinated fibers, decreasing their ability to remain healthy, and leading to demyelination.

In another paper on CPM, the authors relate cases associated with burns and serum hyperosmolality (McKee, A. et al., 1988). This was a retrospective study over 16 years looking at CPM in patients dying during treatment for burns. Results showed a frequency of 10 CPM patients out of 139 burn deaths compared to 10 cases out of 3528 autopsy cases in the general population. Data presented suggest a causal relation between a prolonged episode of extreme serum hyperosmolality and CPM. Evidence sited was hyperosmolality was present in all burn CPM patients and seemed to correspond with the onset of CPM. Other published cases had hyponatremia and/or hyperosmolality.

The sequence of events could be that hyponatremia causes brain swelling. The brain compensates by losing intracellular particles. Water is then lost, reducing swelling. The brain is exposed to a relatively hypotonic insult. Because of hyponatremia and hypoosmolarity, burn patients are 25 times more likely to develop CPM than non-burn populations.

Alcoholic Cerebellar Degeneration

A very detailed and comprehensive study of alcoholic cerebellar degeneration (ACD) appeared in 1959 (Victor et al., 1959). In this study, 50 patients were followed over many years. This group of alcoholic patients had a very specific and uniform collection of symptoms, which were limited to gait ataxia, and ataxia to the lower extremities. This ataxia evolved over a short period of time and was preceded and followed by relative stability. Cerebral lesions were largely located in the cerebellar vermis.

Fifty patients with ACD were followed. There was little doubt that each patient had a long previous history of alcohol consumption, exceeding 10 years duration in most cases. Most had disruptive social lives, and many had DTs, and hallucinations. Careful history and physical examination determined that severe nutritional compromises had occurred in most patients. More than half presented with physical evidence of nutritional deficiencies, such as loss of muscle bulk. Almost half of the patients had liver disease, usually cirrhosis. Eleven of the patients no signs or had histories of poor nutrition.

The average age of onset of cerebellar symptoms was 46. The range was 31–63 years of age. Over 60% had an onset of symptoms before the age of 50. The initial complaint of all patients, except two, was ataxia. This was variously described as weakness, staggering, unsteadiness, or loss of balance. Signs upon exam were nearly all ataxia of gait, and incoordination of legs. Symptoms such as nystagmus were either mild or absent. Besides walking, leg movement disorders were present in all but three cases. The leg symptoms were characteristic cerebellum incoordination, which was best demonstrated on heel to shin testing. Upper extremity coordination involvement was minimal, with several cases showing no effect. A few patients had a rhythmical tremor which appeared with certain postures, but was not the usual cerebellar intention tremor. Cerebellar catalepsy was absent in these patients.

Blood analysis, blood chemistries, and urinalysis were all normal, except if there was a secondary disease such as cirrhosis. CSF was mostly normal in all cases examined. Pneumoencephalograms were only obtained in three patients, but each showed a dilatation of the fourth ventricle and cisterna magna and a small cerebellum as compared to normal. Decreased intellectual capability was only present in three cases in which there were earlier symptoms of Wernicke–Korsakoff syndrome.

Alcoholic cerebellar degeneration in this study had a variable course. In about half (23) of the cases, the symptoms of ACD evolved rapidly, with the severity of symptoms occurring in a few weeks to months. After this worsening period, the progression stabilized. The stabilization seemed to correlate with improved nutrition and abstinence.

In another group of patients, progression of symptoms continued to progressively evolve over many years. Unless cessation of drinking and improved nutrition occurred, these patients continued to get worse. A few of these patients did stop drinking and eating poorly, and their deterioration plateaued. In some cases the period of deterioration lasted as long as 15 years.

A third group showed a relatively stable period which was suddenly characterized by a rapid onset of deterioration. This group then showed plateauing in the advance of symptoms. The rapid advance in symptoms could be measured in weeks to months. An attempt was made to correlate the increase in severity of symptoms to drinking or poor nutrition. In most cases, drinking had been more or less steady preceding the rapid worsening of symptoms. In a few cases, there had been a decrease in alcohol consumption prior to an increase in symptom severity. Plateauing of symptoms was frequently associated with decreased drinking and improved nutrition. Only six patients showed stabilization of symptoms without concurrent improvement in nutrition and decreased alcohol consumption. When the ataxia became established, even cessation of drinking and improved nutrition did not improve symptoms. Only a few patients had any improvement, and this was thought to be due to long-term better health and retraining in walking. This no doubt reflects the more or less permanency of the cerebellar lesion.

Neuropathological examination was possible in 11 ACD patients described in the above study. Grossly, brains from these ACD patients had a more or less normal appearance. Coronal sectioning showed occasional dilatation of the lateral ventricles, but otherwise seemed normal, except for the cerebellum. In some cases, cerebella showed a separation of the sulci and shrinkage of the vermis. Microscopically, the culmen seemed to be most severely affected in most cases. In this region, there were essentially no Purkinje cells. Astrocytic proliferation was present in the Purkinje cell layer in the affected cerebellar area. The molecular layer was only 1/2–1/3 the thickness of unaffected areas. Other areas affected included the flocculi, anterior areas of the anterior lobes, and dorsal paraflocculus. Here too, Purkinje cells were absent and there was astrocytic hyperplasia (Bergmann's astrocytes). The molecular layer in these affected areas was also reduced in thickness, with accompanying astrocyte proliferation. In the affected areas, many fibers were gone, as well as Purkinje dendrites. It was noted that the transition from "affected" to "non-affected" was sharp, occurring over a linear span of a couple of cells. The occasional Purkinje cell in the affected areas was pale and shrunken.

In areas in which destruction was somewhat less, there was still a reduction in Purkinje cell numbers and the remaining cells were pale and small. Even in these areas, the molecular layer was thinned and granular cells were reduced in number. A reduction in axons and dendrites was seen in affected areas, and myelinated fibers were pale.

Deep cerebellar nuclei were also affected. Thus, the dentate, globose, emboliform, and fastigial nuclei were usually described as having a loss of neurons and shrinkage of those present. There was a concomitant proliferation of microglial cells and astrocytes in these nuclei. Lipofuscin deposits were present in remaining neurons. Fibers in the deep nuclei were decreased in numbers.

Certain other areas of the brain usually showed some involvement. For example, the corpus callosum was thinned in some areas as much as 75%. Very mild pathology was seen in a few instances in the cerebral cortex. The olives and superior vestibular nuclei were involved in a few cases. These areas showed some reduced numbers of neurons and occasional inclusions in remaining neurons. Myelin loss was noted in a couple instances in spinal cord columns. The vagus nuclei and the nucleus pontis showed minor involvement in a couple cases. Fibrose gliosis was present in some areas; however, the mammillary bodies appeared normal.

The authors of this paper undertook a Purkinje cell count in ACD patients as compared to control patients dying of unrelated causes. Four patients were controls; two patients were alcoholic but had no neurological symptoms. Results from controls were not significantly different from each other. Purkinje cell counts in these controls ranged from 3.5 to 7.0 cells per millimeter. Counts from patients with ACD showed, in highly affected cerebellar areas, less than 2 cells per millimeter. These numbers were of Purkinje cell presence or absence and did not consider the size of cells or size of the shrunken area of the cerebellum. If the cerebellar area was itself shrunken, that would falsely elevate the number of Purkinje cells per millimeter. Some control subjects were over 70 years of age and may have had minor age related decreases in Purkinje cell numbers, but regardless, there was significant reduction in cells in ACD patients.

The authors attempted to correlate clinical findings in each case to neuropathological findings at autopsy. In some cases, lesions were relatively mild, and patients showed no signs of cerebellar symptoms. In other cases, where symptoms were severe, some had widespread cerebellar lesions and others had lesions restricted to the superior vermis and anterior lobes only. This is interpreted to mean that the symptoms are generated by lesions of the superior vermis and anterior lobes. The neuropathological basis for explaining both nystagmus and speech difficulties could not be determined. The localization of symptoms to the above structures is not in complete agreement with the functions of these cerebellar areas as determined by other authors (Dow, 1938). Certainly the highly specific nature of the lesions, as well as being bilateral, gives credence to the assigning of function by the authors.

More recent studies of the functional organization of the cerebellum (Henneman, 1956; Dow and Moruzzi, 1958) have presented data more in keeping with the clinico-pathological results. These physiological studies involve recording and mapping evoked potentials from cerebellar areas, as well as defining motor effects elicited by the electrical stimulation of similar cerebellar sites. This has resulted in a map of topographical localization of represented body areas on the cerebellum. This representation of body sites on the cerebellum correlates well with similar results described in the ACD patients. For example, experimental physiological/anatomical results show that cerebellar areas shown to be involved in lower extremity function are precisely those lesioned by ACD. This could easily explain the most prominent clinical characteristic, namely ataxia of gait, and leg incoordination. While the culmen was almost always involved, the posterior area of the anterior lobes was infrequently involved, thus the sparing of arm involvement. Head and neck involvement was not seen clinically, and the declive was spared neuropathologically. The site in the cerebellum responsible for nystagmus could not be determined.

The authors state that in terms of etiology, many potential causes are present. Several patients had co-existing diseases such as carcinoma, but serious abuse of alcohol occurred in all patients. In some patients, however, the cerebellar symptoms occurred several years after abstinence started. Some of the abstinent patients were under close supervision, and there was no drinking. This renders the alcohol toxic possibility as an etiologic factor remote.

Another possibility is undernutrition. In fact, 75% of the patients in this study were undernourished. Again, however, several patients had had long periods of normal nutrition. Scant few animal studies on nutritionally induced lesions in the cerebellum exist. Other conditions such as anoxia produce cerebellar lesions, but in different areas than those seen in ACD. Some similar cerebellar lesions are seen in hyperthermia, but the overall picture is of diffuse cerebellar involvement, not the highly focal lesions seen in ACD. Other possible causes, including toxins such as lead, methyl mercury, thiopene, DDT, carbon tetrachloride, nitrogen chloride, manganese, dilantin, and injection of eosinophils, independently can produce cerebellar lesions. For various reasons, such as location and nature of both lesions and symptom production, all of these potential contributors can be excluded.

The authors state that it is speculative to suggest that the cerebellum in these patients is damaged or weakened. The idea is that over time, the cerebellum was brought to a condition such that it was "on the edge" and vulnerable. Arguing against this concept is the fact that there never was a mild or "preclinical" cerebellar state described. There are no childhood cases which might have predicted later development of ACD.

There have, however, been descriptions of a biochemical feature of the cerebellum which might help explain this concept. In most brain regions, there is an overabundance of enzyme capability. Thus, enzymes such as pyruvate dehydrogenase may be present in 2–4 times the amount needed for normal function. The cerebellum, however, seems unique in this regard. Studies have been described (Reynolds and Blass, 1976) showing only a small excess of activity of pyruvate dehydrogenase in the cerebellum over what is required. The idea has evolved that such a small reserve capacity might place the cerebellum in jeopardy more quickly than other brain regions. This unique biochemistry could easily, over many years, be compromised to the point that a small challenge, even after periods of abstinence, could be precipitously detrimental. This concept, hard to prove, awaits verification. It could however be the basis for the rapid deterioration of symptoms in ACD patients.

Marchiafava–Bignami Disease

Marchiafava–Bignami Disease (MBD) was first described in Italy and was thought to only occur in Italian males drinking red wine (Marchiafava, E. and Bignami, A., 1903). Soon it was found that MBD was a disorder primarily seen in alcoholics regardless of ethnic background. MBD is a relatively rare disorder, with only about 300 cases reported in the literature, and nearly all have been chronic alcoholics. The pathological feature which distinguishes this disorder is the symmetric demyelination of the corpus callosum. Lesions may infrequently extend laterally and may involve the cerebral cortex.

In some very rare instances, complete recovery from MBD may occur (Tobita, M. et al., 1997). In this case, T2-weighted MRI showed a high intensity of the splenium and central corpus callosum. The patient was treated with doses of vitamin B complex and a complete recovery was seen 3 weeks later. At this time, MRI showed a nearly complete resolution of the T2-weighted high intensity seen earlier. The authors state that this was an acute early example of MBD, quickly treated with B complex vitamins.

While most cases of MBD are diagnosed at autopsy, increasingly a diagnosis is made from clinical presentations and MRI. In a recent report (Arbelaez, A. et al., 2003), two cases of MBD are reported in alcoholics who were diagnosed by both clinical criteria and MRI. Both patients were chronic alcoholics, one consuming 1–2 bottles of wine per day for years. Clinical evaluation showed malnourishment, short- and long-term memory loss, agraphia, ataxia, and a wide-based gait. Laboratory data included hypoglycemia and decreased serum levels of thiamine. MRI in both patients showed high T2 signal intensities in the corpus callosum, and in one case extending into the white matter of the cortical frontal lobes. In each case, vitamin B complex was initiated, which resulted in marked improvement in symptoms.

The authors point out that MBD is a disease almost exclusively of alcoholic and malnourished patients and is thought to be a B complex (thiamine) related disorder. This disease usually affects the body of the corpus callosum at first and then can reach the genu and splenium. Other white matter areas may be affected in later stages. The corpus callosum lesion may be layered, then show cavitation. During acute phases, macrophages infiltrate. Demyelination may lead to loss of axons.

Patients with MBD may also have Wernicke's disease, or even central pontine myelination. Chronic MBD is characterized by dementia, and patients may live many years. There may be a connection in some cases to hyponatremia, the correction of which can lead to myelinolysis. Diagnostic differentiation between MBD and Wernicke's disease is relatively easy due to the involvement of the mammillary bodies in Wernicke's disease.

Another case of MBD showed the usual involvement of the corpus callosum, as well as symmetrical lesions in the putamen (Hayashi, T. et al., 2002). The patient was a 65-year-old male who was a chronic alcoholic. When admitted, he was comatose and had increased muscle tonis. Laboratory data revealed a decrease in serum thiamine. At this point, thiamine treatment was started and in 24 hours improvement began. An MRI on admission showed swelling with hypointensity on the T1-weighted images from the corpus callosum. The corpus callosum and cerebral white matter showed hyperintensity on the T2-weighted image. In addition, the posterior part of the putamina on both sides was involved with hypointensity on T1 and hyperintensity on T2 images. Fifteen days later (and after 2 weeks of thiamine therapy) the corpus callosal swelling was diminished, but cystic lesions had appeared. The lesions in the putamina were almost gone, and the cerebral white matter lesions were completely absent.

The authors state that this represents the first report of lesions in the putamin completely resolving. This case is also unique in that although the patient had coma, a predictor of poor outcome, he recovered. The authors point out that in MBD cases in which thiamine deficiency can be proven via serum thiamine testing, the deficiency may be reversible with thiamine therapy. In this case, the above was true, and the patient with MBD and comatose recovered. Thiamine treatment might well be initiated in MBD patients when the diagnosis is made.

Clinical symptoms might be rather clear cut as in the above case, or more complicated (Ferracci, F. et al., 1999). In this case, a confirmed alcoholic was admitted with altered speech and gait. MRI shortly after admission showed hypointensive T1-weighted, and hyperintensive T2-weighted images from the anterior/central areas of the corpus callosum. Similar lesions were also seen in the centrum semiovale around the lateral ventricles.

Clinically, the patient showed drowsiness, altered gait, absent ankle and plantar reflexes, lack of speech, loss of reading ability, and loss of writing ability, especially with the non-dominant hand. After the initial acute phase had passed, the patient was mostly mute. Mutism has been described previously in cases of MBD. Features of a hemispheric disconnect due to lesions in the corpus callosum manifest as alexia, as in this case. The agraphia shown by this patient was worst in the non-dominant hand. These symptoms can be explained by the results from MRI.

In another published paper (Demaerel, P. and van Paesshen, W., 2004), a 45-year-old man who was a chronic alcoholic lapsed into unconsciousness. Clinical exam showed peripheral nerve sensory neuropathy, dyspraxia, dysarrthria, and disorientation. MRI showed diffuse swelling of the corpus callosum as well as some degeneration of the cerebral cortex and cerebellum. Laboratory tests showed a low level of serum thiamine. IV thiamine was administered (100 mg/day), with resultant improvement in the clinical picture.

Yet another case of MBD involving successful treatment with thiamine has recently been published (Kuo, C. et al., 2004). This case involved a 39-year-old chronic alcoholic with a history of poor eating. Two weeks before admission, the patient had slurred speech and alterations in mentation. Examination revealed the patient was drowsy, had a wide-based gait, showed ataxia, and had mild bilateral hyperreflexia. MRI of the brain showed hypointensive T1-weighted images and hyperintensive T2-weighted images in the genu to the splenium of the corpus callosum. Diffuse swelling was a feature of this lesion. The cerebellum, cerebral cortex, and centrum semiovale appeared normal.

A diagnosis of MBD was made, and vitamin B complex plus thiamine was initiated. During 12 days of hospitalization, his clinical signs and symptoms improved, and he was discharged. The authors state that imaging studies can easily distinguish between Wernicke's disease and MBD, even though the clinical symptoms and histories are similar. The edema seen in acute onset MBD abates, and is replaced by demyelination and necrosis. The authors emphasize that with imaging techniques, the diagnosis can be made, and vitamin B complex and thiamine instituted. This aggressive and rapid treatment with thiamine has been responsible for stopping the progression of MBD, and reversing some symptoms, as occurred in the case presented.

The utility of MRI as a key diagnostic technique has been investigated (Meneegon, P. et al., 2005) This was a retrospective study in which six patients were identified as having MBD. Five of the six were chronic alcoholics and all had MRI during the acute phase. Results of MRI showed hyperintensity of the corpus callosum. Cerebral cortex lesions were found in the fronto-parietal cortex in threee of the six patients. Of the three who did not have cortical lesions, improvement or full recovery occurred. Of the three with cortical lesions, two died and one was left with a severe cognitive impairment.

The authors state that low apparent diffusion coefficient and lesions in the cerebral cortex are associated with a poor prognosis. This is the benefit to MRI in that it can serve to predict outcome. Immediately after diagnosis, thiamine treatment was initiated, which was beneficial in some of the patients. The authors state that diffusion weighted MRI enables an improved capability for diagnosis in MBD. If diagnosis can be made accurately at an earlier stage, then treatment regimes have a better opportunity for success.

In keeping with the above case is a description of MBD in a chronic alcoholic (Haas, L. et al., 2006). This patient was found unconscious and with dysarthria. Examination showed the patient was comatose, with symmetrical pupil reactions. While slight diverging strabismus was present, most other findings were normal. His serum thiamine was low. CT scan showed no abnormalities. An MRI showed high intensity T2 lesions in the corpus callosum and internal capsule. These images represented edema and demyelination.

The authors state that if MBD is mild or has a recent onset, CT scans may be negative. The clinical manifestations may be difficult to interpret as regards a diagnosis. Before MRI, the diagnosis was made at autopsy. Use of MRI can reveal lesions in the corpus callosum, which almost always indicates a diagnosis of MBD. Although an uncommon occurrence in chronic alcoholics, making the diagnosis with MRI at an early stage will justify thiamine treatment. In this patient, clinical treatment was noted 2 days after thiamine treatment (100 mg/day) was initiated.

Occasionally MBD can be associated with "overlapping" findings of some other alcohol associated thiamine deficient disorder. In this case, lesions similar to those in CPM are described (Goswami, P. et al., 2008). This patient was admitted in a stuporous state, and had a 15-year history of alcoholism. The patient had mild hyponatremia and decreased serum thiamine levels. MRI findings showed areas of T1 hypointense and T2 hyperintensities in the corpus callosum. These lesions were in the central areas of the corpus callosum. Also, the sagittal image showed a T2 hyperintense area in the pons. Based on the MRI a diagnosis of MBD was made, and thiamine treatment initiated, and a gradual correction of the hyponatremia was begun. Improvement in symptoms was noted within a few days, and a fairly good recovery was achieved.

The authors note that the MRI was characteristic of MBD in that the lesion in the corpus callosum showed the main lesion located in the middle layer, with the dorsal and ventral layers spared. This produces a "sandwich"-like appearance. The body of the corpus callosum is usually affected first, followed by the genu and splenium. Over time these lesions may spread to the putamin and cerebral cortex.

While patients were formerly diagnosed at autopsy, MRI has speeded diagnosis and the initiation of thiamine therapy, which may be beneficial. Many patients with an acute form of MBD may become comatose and die. At the other end are more mild, chronic cases which may be amenable to treatment and survive many years. The use of MRI enables the physician to estimate the extent of the disease. Lesions limited to the body of the corpus callosum may be early, whereas lesions extending to the cerebral cortex and putamin may represent a longer lesion process and a less favorable outcome.

The finding in this case of CPM is particularly rare, but points out that one must be alert to variations in not only clinical presentations, but also in lesion distribution.

In another case report (Fukumoto, J. and Suzuki, T., 2007), a 60-year-old alcoholic was admitted in a stuporous condition. Examination showed polyneuropathy with diminished deep tendon reflexes, hypertonic limbs, distal muscle atrophy, gait disturbances, and amnesia. MRI showed T1 hypointensities in the corpus callosum and T2 hyperintensities in the same areas. There were no other lesions noted in the cerebellum or brain stem. There was significant cortical atrophy. Laboratory data showed low serum thiamine levels. The patient also had signs of disconnection syndrome, namely apraxia, acalculia, and dyslexia. Thiamine, vitamin B complex, and mianserin hydrochloride (for depression) therapies were started.

A gradual recovery of many earlier symptoms occurred. Thus, motility improved, gait deficiencies decreased, dysarthria disappeared, and socialization improved. By 40 days, apraxia diminished. By day 60, the corpus callosum lesions seen earlier on MRI had clearly decreased in size. Later still, his disconnection disorder had completely resolved.

The authors note that MBD can be an acute inflammatory disorder in which the splenium of the corpus callosum is affected. The disorder occurs almost always in alcoholics, and is linked to disordered thiamine metabolism. Usually patients are subject to one of three types of MBD: acute, subacute, and chronic. Both acute and subacute tend to have poor outcomes, but the chronic form may be amenable to thiamine treatment. The authors note that in a study where asymptomatic chronic alcoholics had MRIs, two-thirds showed a reduced size of the corpus callosum (Harris, C. et al., 1993). It is not known whether MBD represents an end stage of years of deterioration of the corpus callosum, or represents an increased susceptibility of some patients to corpus callosum damage. An acute phase might argue in favor of the latter.

Hemorrhages are probably a feature of the lesions since hemosiderin is characteristically found in the walls of the callosal cystic lesions. Such hemorrhagic lesions are also seen in lesions in Wernicke's disease

Mianserin hydrochloride was used as an antidepressant drug and also for its ability to decrease delirium. The authors suggest that the mental features of MBD should be treated as well as using vitamin B complex and thiamine to treat the physical aspects of MBD.

Thiamine treatment does not always work. In a case report (Kim, M. et al., 2007), a patient was admitted stuporous and had had seizures, altered mental status, and had consumed large quantities of alcohol for at least 10 years. MRI showed T2-weighted hyperintensities, which were localized to the splenium and body of the corpus callosum. Lesions were also found in the frontal cerebral cortex.

This patient was diagnosed and placed on vitamin B complex treatment plus 1000 mg/day of thiamine and corticosteroid delivered IV. In spite of this, the patient remained stuporous. MRI at 15 days showed decreased signal density in the cerebral cortex lesions, and corpus callosum. Twelve weeks later, MRI showed corpus callosum atrophy, with multiple lesions in the splenium, body, and genu. The patient remained in a semi-vegetative state.

The authors note that the diagnosis can be difficult, but the lesions in the corpus callosum and lack of lesions in the mammillary bodies were strongly suggestive of MBD. This patient was unique in that he had diffuse frontal cortex lesions. These have been seen before and their generation is unclear. They might be from a long time existence of corpus callosum lesions, or have arisen due to some other condition such as hyponatremia and its rapid treatment.

An unusual case as regards symptoms was presented (Hirayama, K. et al., 2008). This was a chronic alcoholic patient who showed on MRI symmetric demyelination and necrosis in the genu, body, and anterior splenium of the corpus callosum. The unusual symptom related to the coordination syndrome seen in most cases of MBD. In this patient, the dyspraxia took a form in which both of the patient's hands could cooperate in a sequence of bimanual actions. For example, his right hand might start an action with the left hand cooperating. Striking was that once the action was completed the left hand would commence undoing the action in an antagonistic fashion. In this antagonistic action, the right hand cooperated with the now "dominant" left hand. These "back and forth" actions continued until the patient was forcibly stopped. This occurred only in bimanual tasks and not at all in voluntary actions. They only occurred in response to verbal commands or in actions imitating someone else.

The authors speculate that this antagonistic action could result by conflict between the two hemispheres due to inter hemispheric disinhibition caused by the partial demyelination of the corpus callosum. Any bimanual coordination is still possible because the corpus callosum is not completely disrupted.

As a sort of plea for MRI and early diagnosis is the report of two cases in which the presentation was an acute onset coma (Vazquez, C. et al., 2008). MRI revealed significant lesions in the corpus callosum. As many previous studies have shown, an acute onset with extensive corpus callosum (and possible involvement of the frontal cerebral cortex) is an ominous feature. These are most frequently associated with death, whether or not thiamine therapy is initiated.

MBD has gone from a disorder solely diagnosed at autopsy and not treated to one frequently diagnosed by MRI and amenable to treatment. This behooves the clinician to be alert to the possibility of MBD in chronic alcoholics. When the diagnosis is made, thiamine treatment should be immediately started. Vitamin B complex treatment should also be started. The possibility that quick treatment might lead to improvement – even complete resolution – of some or all symptoms and signs places MBD into a classification of "treatable," which was not considered before modern imaging permitted the rapid ability to diagnose this disorder.

Leigh's Disease

The disease subacute necrotizing encephalomyelopathy (SNE) was first described in 1951 by Denis Leigh and therefore is also called Leigh's disease (Leigh, 1951). This disorder, although afflicting mainly children, has many features similar to those of Wernicke's disease. These neuropathological features were so striking as to have been noted in the first case description by Dr. Leigh. Many studies have evolved over the years describing various biochemical features of Leigh's disease. Most recently, genetic and mitochondrial alterations in brain, blood, and tissue of SNE patients have been described. This chapter will examine results from these studies in a chronological order and try to reach some valid conclusions and note where information is lacking.

Denis Leigh described the first case of SNE in 1951. This was a case of a 7-month, 3-week-old boy admitted to the hospital with pupils unreactive to light and limb spasticity, and he appeared to be deaf (see Table 1). Over a period of 3 days, the patient became rapidly worse, lapsed into a coma, and died.

An autopsy was performed nearly 3 days later and was unremarkable except for CNS findings (see Table 2). Upon gross examination of the brain, it was noted that mid-thalamic regions showed areas of brownish discoloration in the dorsomedial nuclei, nucleus submedius, substantia nigra, and the pulvinar. In the midbrain, similar lesions were noted in the entire tegmentum. In the pons, lesions were found centrally, and also in the trigeminal motor nuclei. The medulla showed the same brownish lesions in the inferior olivary nuclei, and in the central gray. The posterior columns were similarly affected in the spinal cord. The lesion became less conspicuous at lower spinal cord levels. There was a striking bilaterally symmetrical characteristic to these neuropathologic lesions.

When sections were stained with Perdrau's silver stain, the regions which appeared to have a brownish color grossly were all sites of proliferation of small blood vessels, including arterioles, pre-capillaries, capillaries, and venules (see Table 3). The proliferation was not associated with hemorrhage or wall thickening. Neurons in close proximity to these vascular alterations were damaged. Neuronal damage included cell body swelling, chromatolysis, and nuclear swelling. Microglia were increased and were concentrated at the periphery of the lesion. Oligodendrocytes were reduced in number. Sections stained with Heidenhain's stain

D.W. McCandless, *Thiamine Deficiency and Associated Clinical Disorders*, Contemporary Clinical Neuroscience, DOI 10.1007/978-1-60761-311-4_7,

for myelin showed widespread demyelination in affected areas. Microscopic damage followed that seen on gross inspection. In the midbrain, the oculomotor nuclei were severely damaged. Striking again was the bilateral symmetry. Also noteworthy was the complete sparing of the mammilary bodies. At the level of the medulla, the destructive process was limited to the inferior olives only. In the spinal cord, microscopic changes were limited to the dorsal columns. These changes diminished with descending levels of the spinal cord.

In the discussion of this case, Dr. Leigh comments on the similarity of the lesion in this case to that seen in Wernicke's disease. He also comments on the striking bilateral symmetry of the lesions and on the highly localized, punctate nature of the lesion. In considering a diagnosis, the author mentions some form of poisoning (arsenic), equine encephalitis, and subacute necrotizing myelitis as possible diagnoses, but ultimately he rules them out.

Finally, the author is striken by the similarity of the lesions in this case to those seen in Wernicke's encephalopathy. These similarities include a vascular response, nerve cell damage, and proliferation of microglia and astrocytes. There are differences in regional sites between this case and Wernicke's disease, most notably, the mammillary bodies, damaged in Wernicke's disease, were spared in this case. Also, spinal cord lesions, as seen in this case, are not commonly noted in Wernicke's disease.

Leigh points out in his 1951 paper, that very similar CNS lesions were produced in earlier studies by administering quinoline compounds to monkeys (Richter, 1949). The lesions in these experimental monkeys were vascular in nature, highly specific in location, and bilaterally symmetrical. The suggestion was offered that this pattern was closely matched to lesions seen in Wernicke's disease. In this first case description of SNE, Leigh considers the diagnostic options of an unknown toxin or virus, but seems to favor an infantile form of Wernicke's disease as the possible cause of this new encephalopathy.

Another early comprehensive description of Leigh's disease Richter (1957) details symptoms and pathological changes in three cases. All three had similar symptoms, which began between 7 and 13 months. All three died after periods of illness ranging from 4 to 14 months. Symptoms in these three cases were quite similar and included reduced food intake, diminished to no response to light and sound stimuli, retardation, seizures, spasticity, stupor, and death (see Table 1).

Upon gross examination of the brains of these three cases, many similarities were noted, compared to the earlier cases (see Table 2). Areas showing gray-red discoloration and a softer texture included the caudate nucleus, putamen, basal ganglia, striatum, substantia nigra, tegmentun of the pons, and the medulla. There appeared to be a decrease in the relatively sharp demarcation between brain regions. The bilateral symmetry of these gross lesions was noted.

Microscopically, lesions were generally similar to those of earlier studies (see Table 3). These consisted in the first case of areas of degeneration of the periaqueductal region and substantia nigra. The degeneration consisted of necrosis involving the ground substance, as indicated by incomplete staining. There were notable vascular changes which included an increase in dilated pre-capillaries and capillaries.

In these areas, there was some limited neuronyl degeneration, but the neurons were largely spared. There was demyelination and astrocytic proliferation. In the pons, case 1 showed extensive demyelination throughout the tegmentum, and in the medulla, there was increased vascularity, gliosis, and demyelination throughout the reticular formation.

In the second case of Richter, many histological sections took the various stains poorly, signaling degeneration of ground substance. In lesioned regions, there was the consistent finding of increased prominence and/or numbers of pre-capillaries and capillaries. In affected areas such as the occulomotor nuclei, the neurons were normal, but there was significant astrocytic proliferation throughout the nucleus. The trochlear nuclei were similarly affected. The myelin pattern in the brainstem was affected (demyelinated) in the tegmentum and substantia nigra. The pontine tegmentum showed vascular proliferation in the locus ceruleus, caudate, putamen, globus pallidus, and subthalamic nuclei.

In Richter's case 3, the tegmentum of the brainstem was spongy on sectioning, and lesions were found in the trochlear nucleus. Neurons in this case were relatively spared, but there was capillary and pre-capillary proliferation (prominence). Lesions in the substantia nigra were the same, i.e., neurons were relatively spared. Lesions in the pons also consisted of pre-capillary and capillary prominence, with gliosis prevalent. In the caudate nucleus, most neurons were gone and the spaces were occupied by a framework of proliferating capillaries. Microscopically, lesions were noted in the putamen and the thalamus. Myelinated fibers in these areas were greatly reduced or absent.

There were striking similarities in Richter's cases as compared to previously described cases (Tables 1, 2, and 3). First, was the bilateral symmetry of the lesions, noted in all cases. Also noted were the highly specific and localized nature of the destruction. Finally, in each case of SNE was the similarity of cerebral lesions to those seen in cases of Wernicke's disease, except for the sparing of the mammillary bodies in SNE. Richter speculates that given the relative vulnerability of the newborn brain, it is odd that in SNE, the mammillary bodies are spared, while almost always involved in Wernicke's disease. Richter suggests, however, that a clear pathological relation exists between Leigh's disease and Wernicke's disease. Richter, at this early time, suggests that SNE, unlike Wernicke's disease, may have a genetic base. Although Wernicke's disease cases usually show some level of nutritional disorder, Leigh's cases may have "feeding difficulties," but overall, nutritional status is about normal. From this, Richter suggested that there might be a "poisoning" of metabolic enzymes, or that there could be an inherited enzymatic defect responsible for the clinical disorder of Leigh's disease.

In his discussion and conclusions, Richter noted the similarity of pathological findings in his three cases to those described by Feigin and Wolf (1954). Feigin and Wolf (see Tables 1–3) noted similarities of Leigh's disease lesions to those seen in Wernicke's disease. While Feigin and Wolf pointed out the feeding problems in cases of Leigh's disease, these authors felt that this was a result of the disease as opposed to a cause or contributor to the disorder. The fact that these early descriptions of Leigh's disease were so similar as regards to pathological changes, leads to the conclusion that they collectively form an identifiable pathological entity which is unique from other general infantile degenerative disorders, and from Wernicke's disease.

There followed papers describing cases of Leigh's disease in combination with other symptoms and findings not ordinarily associated with SNE (Kamoshita et al., 1970). These authors found infantile spasms and hyperarrhythmia, and an abnormal EEG. The uniqueness of these combined findings is discussed, but no conclusions are drawn. In other studies, the association of Leigh's disease and renal and arterial lesions are described (Crome, 1970), and ocular findings have also been examined (Grover et al., 1970).

In 1967, Yashon and Jane (1967) described the 21st–23rd cases of Leigh's disease. Symptoms in these three cases were overall similar to those described in previous cases (Table 1).

SNE SYMPTOMS

	VISUAL	HEARING	SPASTICITY	STUPOR/COMA	SEIZURE ACTIVITY
D. LEIGH 1 CASE	+	+	+	+	
R. RICHTER 3 CASES	+	+	+	+	
YASHON & JANE 3 CASES	+		+	+	+
FEIGIN & WOLF 3 CASES	+		+	+	+
DEVIO, ET.AL 1 CASE	+	+	+		+

Table 1

However these authors point out that varying degrees of pathologic involvement influence the expression of symptoms in each individual case. Seizures were present in one of the three cases described by Yashon and Jane and in only two previously reported cases. Malnutrition, present in some cases, was absent in one of these three cases. In the three cases of Yashon and Jane, there was no evidence of any genetic link, as was noted in previous studies, which had shown a high incidence of perinatal deaths in family histories.

Regarding the cerebral lesions, Yashon and Jane (1967) point out that prominent capillary vascularity and breakdown of ground substance, giving a loose appearance, are so characteristic that classification of a case of Leigh's disease without these features would be impossible. Again, Yashon and Jane comment on the symmetry of lesions, the relative sparing of neurons in lesion areas, and the striking similarity to the lesions of Wernicke's disease. Unusual in these cases was the finding of involvement in the mammillary bodies. Mammillary body lesions are almost always present in Wernicke's disease, and involvement of mammillary bodies in one case of Yashon and Jane is only the second report of its kind in the literature in Leigh's disease. Again, in relation to Wernicke's disease is speculation that, because of lesion similarity, some sort of thiamine metabolism alteration must be considered in Leigh's disease. The paper of Yashon and Jane concludes with an interesting discussion about the relation of SNE to other thiamine-related disorders besides Wernicke's disease. For example, the cerebral lesions in Chastek paralysis, which are very similar to those seen in both SNE and Wernicke's disease, are examined. Chastek paralysis (in foxes) was eliminated (reversed) by thiamine treatment. There is speculation of possible enzyme antagonist activity in human cases which might be responsible for the symptoms and lesions (Tables 2 and 3).

SNE GROSS BRAIN FINDINGS

	THALAMUS	PULVINAR	INF. COL.	PONS	OLIVE	MEDULLA CENTRAL GRAY	POSTERIOR COLUMN S.C.	SUBSTANTIA NIGRA	CORTEX	CAUDATE
D. LEIGH 1 CASE	+	+	+	+	+	+	+			
R. RICHTER 3 CASES			+	+	+	+		+	+	+
YASHON & JANE 3 CASES	+			+	+	+	+		+	+
FEIGEN & WOLF 3 CASES			+	+	+	+		+	+	
DEVIO, ET. AL 1 CASE	+	+		+			+	+		+

Table 2

SNE HISTOLOGICAL FINDINGS

	VASCULAR PROLIFERATION	NEURAL DEGENERATION	ACTIVE MICROGLIA	DEMYELINATION	GLIOSIS (ASTROCYTIC)	SOME NEURAL SPARING
D. LEIGH 1 CASE	+	+	+	+	+	
R. RICHTER 3 CASES	+	+	+	+	+	+
YASHON & JANE 3 CASES		+	+	+	+	
FEIGEN & WOLF 3 CASES	+	+	+	+	+	+
DEVIO, ET. AL 1 CASE	+	+	+	+	+	+

Table 3

In 1968, a paper was published in which lactate and pyruvate were studied in a child with supposed Leigh's disease (Hommes et al., 1968). Results showed that both lactate and pyruvate were elevated in this patient, who showed symptoms similar to those of Leigh's disease, and who had three siblings die of symptoms said to be those of Leigh's disease (Clayton et al., 1967). A liver biopsy showed an almost complete absence of pyruvate carboxylase, a gluconeogenic enzyme. Data from this patient were suggestive for a defect in gluconeogenesis. With this in mind, this patient was treated with lipoic acid. As described before (Clayton et al., 1967), lipoic acid had a significant improvement in symptoms, and lactate and pyruvate blood levels dropped. The authors emphasized that this case was suggestive of Leigh's disease; at the time of publication, the patient was alive, and there was no pathological conformation of the diagnosis. This paper did reconfirm possible beneficial effects of lipoic acid treatment.

In 1969, a "factor" was described in the urine of a patient with Leigh's disease which inhibited the enzyme thiamine pyrosphate-adenosine triphosphate phosphotransferase (Pincus et al., 1969). The reaction is as follows: TTP + ATP → TTP + ADP. Postmortem biochemical analysis in Leigh's patients yield TTP values in brain, liver, and kidney near zero as compared to normal control tissue. In these patients, activity levels of pyruvate decarboxylase, alpha-ketoglutarate dehydrogenase, and transketolase (all thiamine-requiring enzymes) were slightly decreased or not changed at all (alpha-ketoglutarate dehydrogenase). Not only urine but blood and spinal fluid also contained the inhibitor.

Because of the striking similarity of cerebral lesions in Wernicke's disease and Leigh's disease, investigators were looking for some alteration in the known neurochemical reactions involving thiamine. In the Pincus et al. paper, the finding of an inhibitor in body fluids of the reaction forming TTP seemed exciting because of the following: (1) this could explain the development of the disorder; (2) this could also explain normal levels in vitro of thiamine-requiring enzymes; and (3) this finding could become the base for a relatively simple urinary test for Leigh's disease. In this case, brain tissue was obtained 45 minutes after death, and samples were analyzed for TTP. Results showed a virtual absence of TTP in brain samples, although TTP was present in liver and kidney. This finding suggested that the inhibitor is only active in brain tissue. As was pointed out by the authors, thiamine monophosphate constitutes 10% of thiamine in the body, TPP about 80%, and TTP about 10%. The physiological role of TTP in brain function is not clear, but may play a role in nerve conduction.

In a subsequent paper (Cooper et al., 1970), description is presented of results of screening samples of urine from cases of possible SNE, as well as normal controls. These studies showed five positive cases of SNE, and three probable cases based on the urine test for inhibitor. There were three false positive tests – two in siblings of SNE patients and one in a case of metachromic leukodystrophy. Results are also described stating that the inhibitor is a labile glycoprotein with a M.W. of around 30,000. The lability makes isolation and characterization difficult. It is stated that several patients diagnosed with SNE were at that time being treated with thiamine and have shown improvement.

In a somewhat later paper (Pincus et al., 1971), the effects of treatment of a case of Leigh's disease with thiamine and thiamine propyl disulfide were described. Initially, treatment with the above compounds was associated with a dramatic improvement in neurological symptoms. Later, however, thiamine levels in CSF dropped, and symptoms became worse. Following death, tissue from the patients' brain showed low thiamine levels, and TTP was absent in samples taken from the pons. The reasons for these results are unclear.

Another interesting paper by Simopoulos et al. (1972) details changes in SNE not heretofore described, and reflects the complexity of this disorder. This paper describes a case of SNE in which the patient displayed many of the symptoms of SNE, including visual symptoms, poor muscular development, and ataxia. The patient was found to be hyperpneic and had hypomagnesemia and hypokalemia. Both blood and urine were analyzed for the inhibitor, and the results were positive. Based on this, and the clinical findings, SNE was suspected. Anemia was an early feature, and a bone marrow aspiration showed pronounced vacuolization. A renal biopsy was normal. Because of the close relation of SNE and Wernicke's disease, a treatment regime of thiamine was tried. Administration of thiamine resulted in a disappearance of the inhibitor, and 3 months after initiating thiamine treatment, a bone marrow aspiration showed that the vacuoles seen previously had disappeared. Upon discontinuation of thiamine treatment, the vacuoles had reappeared in the bone marrow. Symptoms seemed to improve such that the patient was able to attend kindergarten. Her IQ at that time was 90.

Subsequently, the patient had headaches, refused to eat, refused medications, and was readmitted to the hospital. She was dehydrated, emaciated, semi-comatose, and showed opisthotonic posturing. The patient was clearly in a severe relapse, and despite IV thiamine administration died.

Postmortem exam revealed no gross abnormalities in the cerebrum, basal ganglia, brainstem, or cerebellum. The spinal cord had discoloration in the dorsal columns. Microscopically, there was a striking vacuolization in the white matter. Myelin loss was most conspicuous in the dorsal columns. Some vacuolization was seen in subcortical fibers, and mild neuronyl loss was noted. Mild vacuolization was present in the basal ganglia, thalamus, hypothalamus, and mammilary bodies, but there was no neuronyl loss or vascular prominence. This same vacuolization was noted in the reticular formation and periaqueductal gray.

As pointed out by the authors, respiratory alkalosis, hypokalemia, hypomagnesemia, aldosterosemia, and vacuolization of bone marrow have never before been described in Leigh's disease. Regarding brain pathology, the mammillary bodies in this case showed changes rarely seen in other cases of Leigh's disease. In addition, other brain lesions, gross and microscopic, do not duplicate those characteristically seen in Leigh's disease. Even the course of this patient's disease, drawn out over several years, is not consistent with other cases. This patient did have the inhibitor present in both blood and urine.

Thiamine has been previously reported to have beneficial effect in the treatment of megaloblastic anemia and in this case resulted in the decrease in bone marrow vacuoles. Thiamine also resulted in the disappearance of the inhibitor in blood and urine. When thiamine treatment was stopped, these diminished changes returned. The bone marrow alterations, coupled with the presence of the inhibitor, suggest a variant of SNE not yet described. The lack of characteristic clinical course and of neuropathological changes argues in favor of the above hypothesis. This case would probably not been called Leigh's disease 10 years earlier.

In a paper not directly associated with a symptomatic Leigh's patient, results are presented of the measurement of the urinary inhibitor in the parents of seven cases of Leigh's disease (Murphy, J. et al., 1974). The inhibitor of the TPP+ATP → TTP+ADP reaction was similar in parents to that in confirmed Leigh's patients in that (1) thiamine treatment reduced the levels of the inhibitor, (2) the inhibitor becomes inactive after three weeks of frozen storage, and (3) the inhibitor is destroyed by boiling water. Because the inhibitor in parents was found at the same levels as in symptomatic patients, this assay cannot be used to identify patients before onset of symptoms. Also of interest is the finding that parents with inhibitor are asymptomatic.

In another paper (Hommes et al., 1973), the effect of thiamine administration in rat liver mitochondrial pyruvate dehydrogenase activity was examined. In this study, rats were injected IP with thiamine dichloride and after five days, liver mitochondria were prepared and analyzed. Results showed that inhibition of the enzyme by ATP was reduced in treated as compared to non-treated rats. The implication is that thiamine administration "improves" the ability of pyruvate dehydrogenase to metabolize pyruvate, thereby increasing the efficiency of the TCA cycle. This could serve to explain the beneficial effects of thiamine treatment in Leigh's patients.

Further studies of inhibitor characteristics (Murphy et al., 1974) showed that all obligate carriers of Leigh's disease have inhibitor in their urine. Further, it was found that the inhibitor could be stabilized by 50 mM mercaptoethanol. That the inhibitor was not found in an active form in controls suggested that it is in a unique easily measured marker for SNE. The inhibitor has a M.W. of around 37,000. This paper also demonstrated that the inhibitor is specific for the brain reaction of TPP + ATP → TTP + ADP; the inhibitor had no effect on the liver reaction. The question of why do carriers have similar inhibitor activities, yet are asymptomatic is discussed. It is suggested that (1) the level/activity of inhibitor in brain tissue is not the same in patients vs. carriers or (2) that the sensitivity of dietary thiamine in carriers is more "efficient" in mediating the inhibitor effect than it is in patients with overt Leigh's disease (Pincus et al., 1973).

Pincus et al. (1973) published a paper in which thiamine treatment was examined in a large group of 21 SNE patients. Two groups of patients were studied: (1) 12 patients with a secure diagnosis of Leigh's disease, largely by autopsy, and (2) nine patients in a "presumptive" category. Results showed that in the group of 12 patients, 9 died. Why this occurred is discussed in light of the fact that with treatment, several enjoyed periods of remission. The remissions in the treated group of patients were more significant and lasted longer than those of untreated cases from the literature The ultimate result of mortality was discussed. Three reasons are cited to explain why, when treated, 9 of 12 patients died The first possible reason is that treatment was initiated in at least three cases when the patient was in extremis and had already developed irreversible brain structural damage. A second reason could be the development of an adaptive mechanism, which results in the progressive lowering of thiamine levels, despite treatment, and the third possibility is that SNE is not a simple vitamin deficiency, but involves direct inhibition of thiamine-related reactions.

The fact that carriers of the inhibitor are asymptomatic (Murphy et al., 1974) and the aforementioned suggestion of the development of an adaptive mechanism to offset thiamine treatment are interesting. It could be that thiamine or its ester in carriers is somehow more efficient in mediating the inhibitor effect than is the case in patients with overt SNE.

The question of the specificity of the urine inhibitor as a test for Leigh's disease has been carefully examined (Pincus et al., 1974). In this study, the urine of over 500 subjects was tested for the presence of the inhibitor. Subjects were arranged in a variety of categories, including autopsy-proven cases, living cases of Leigh's disease, patients thought to have Leigh's disease, patients once suspected of Leigh's disease, normal controls, etc. Results showed the false positive rate was 6.4%. More than one-third of parents of Leigh's cases tested positive for the inhibitor. In 16 cases of Leigh's disease confirmed at autopsy, or living, with siblings confirmed at autopsy, all had the inhibitor present in their urine. In 42 patients suspected of Leigh's disease, but later proved to have some other disorder, 39 tested negative for the inhibitor. In this study, no false negatives were found in untreated patients when the urine samples were handled properly. The authors are careful to state that the test is more definitive in ruling out the diagnosis of Leigh's disease than it is in establishing it due to the 6.4% false positive rate.

In a somewhat more thorough autopsy examination report, pathology in other organs besides the brain have been described (Crosby and Chou, 1974). This was an autopsy-proven case of Leigh's disease, with a somewhat unusual course in that the patient had suffered a traumatic brain injury, had a drawn out course of symptoms (over 4 years), and died at the age of 10. At autopsy, the brain showed typical gray brain discolorations in the globus pallidus, hypothalamus, substantia nigra, tegmentum of the medulla/pons, and lesions were noted in the spinal cord. Microscopic examination was characteristic of Leigh's disease. The main feature was a proliferation and hyperplasia of capillaries in affected areas, and demyelination. In most areas, neurons were spared.

In addition to the cerebral changes, light and electron microscopy were performed on biopsy samples and on samples taken at autopsy. Findings included the presence in muscle biopsy samples of red staining (trichrome stain) clumps of material in about 5% of fibers. These clumps were called "ragged-red fibers." Most other features of the muscle fibers were within the normal range, except mitochondria, which appeared enlarged. They contained rectangular paracrystalline arrays located in the crystal membranes. There were also intramitochondrial dense bodies. These characteristic mitochondrial changes are associated with several mitochondrial disorders. In Leigh's disease, biochemical features such as elevated lactate and pyruvate, decreased pyruvate dehydrogenase activity, and altered amino acid metabolism are all noted. The authors of this paper speculate that previous autopsy examination has focused on the brain, and perhaps missed muscle and heart changes seen in this case. The idea is advanced that Leigh's disease may primarily be a mitochondrial disease, with variations, and that further studies should also examine muscle and heart. The existence of cardiomyopathy and mitochondrial myopathy in an autopsy-proven case of Leigh's disease stresses the importance of a full thorough investigation and suggests that some cases of mitochondrial myopathy may actually be Leigh's disease.

In another study, the cerebral levels of thiamine, thiamine monophosphate, thiamine pyrophosphate, and thiamine triphosphate were measured in brain tissue from proven Leigh's cases and non-Leigh's controls (Murphy and Craig, 1975). Results showed that levels of thiamine triphosphate were significantly decreased in Leigh's patients as compared to controls. In addition, thiamine pyrophosphate levels were elevated in Leigh's brains as compared to controls. This suggests that the inhibitor is active in inhibiting the phosphotransferase activity. Brain tissue from Leigh's patients and controls were able to catalyze the reaction ATP + TPP → TTP + ADP in vitro, demonstrating that the enzyme was present, but blocked in Leigh's patients in vivo. The inhibitor did not block the reversal of this reaction. The authors state that these data suggest that the inhibitor plays a major and significant role in the etiology of Leigh's disease. This study does not rule out other possible contributory (mitochondrial) defects. The authors speculate that since both Wernicke's disease and Leigh's have many pathological changes in common, and both have a deficiency in TTP, then TTP must play an important role in normal cerebral metabolism. There is no inhibitor present in Wernicke's patients, yet levels of TTP are low due to overall thiamine deficiency. Thus the depleted levels of TTP result from a different mechanism, yet there are impressive neuropathologic similarities.

In a brief paper published in Lancet (Blass et al., 1976b), the authors measured pyruvate dehydrogenase in fibroblasts from two Leigh's patients. These patients had classic symptoms, as did siblings who died with similar symptoms. And one of the two had autopsy-confirmed Leigh's disease. Results showed a 75% decrease in activity of pyruvate dehydrogenase activity in these two patients' fibroblasts. The point is made that Leigh's disease may have multiple biochemical etiologies and that care should be exercised in defining the specific cause in each case, as the therapy may vary depending on this single characteristic.

In another paper published at about the same time (Murphy, 1976), many cases of Leigh's disease were reviewed. Most cases fell into two groups: (1) those with lactic acidosis, and with a focus on Krebs cycle intermediates, and (2) those in which thiamine and its esters were studied. In this review, in contrast to the previous paper (Blass et al., 1976b), studies were cited in which pyruvate carboxylase was not changed in several Leigh's patients. Murphy states that in all Leigh's patients in whom the inhibitor was tested, it was present. Furthermore, when thiamine phosphate esters were compared in the brains of controls to the levels in Leigh's patients' brains, the levels of thiamine diphosphate were significantly elevated. Further, the levels of thiamine triphosphate were significantly decreased. These changes seem to be specific for brain, as liver levels are not altered in patients. This author speculates that either the increase in thiamine diphosphate or the decrease in TTP is significant and has some as yet undefined role in CNS function.

Because of the supposed involvement of altered thiamine metabolism, numerous therapeutic attempts have been tried over several years. The most common is to administer thiamine in high doses in the form of thiamine disulfide or thiamine tetrahydrofurfuryl disulfide. Initially, there is usually a more or less dramatic improvement in symptoms, weight gain, and a lowering of inhibitor levels in the urine. Later, however, in spite of thiamine administration, the symptoms and inhibitor return, and the downward course continues.

Reasoning that a block in thiamine metabolism relating to its role as a coenzyme could slow the Krebs cycle, leading to decreased energy production, the Krebs cycle was fed from a different site (McCandless and Hodgkin, 1977). In this study, a brother and sister, both with classic Leigh's symptoms, and one with a positive urine inhibitor test were studied. These two Leigh's patients were administered L-glutamine, which was designed to feed the Krebs cycle through glutamate and alpha-ketoglutarate, thereby partially bypassing thiamine-requiring steps. Results from this study were equivocal, and similar to those using thiamine treatment. There was a significant weight gain, and small improvements in symptoms, which lasted several weeks. Then occurred a slow return to the pretreatment condition and continued deterioration. These two patients were far along in their course when this therapy was attempted, and treatment with L-glutamine would be worth trying immediately after suspicion of diagnosis.

In 1978, a paper was published (Devivo et al., 1978) in which a careful study of a Leigh's patient's biochemical changes were studied. In this case, there was also careful postmortem documentation of the neuropathological lesions. A disturbance of the oxidative decarboxylation of pyruvate was suggested by lactic acidosis, hyperglycemia in response to IV alanine, and an elevated flux rate for glucose.

During two hospitalizations, blood lactate levels were elevated. Pyruvate was also elevated, but not as dramatically. Administration of alanine produced a hyperglycemic response. Fibroblast enzyme activities were measured in the patient and parents and were found to be similar to those in blood.

Postmortem measurement of liver and brain pyruvate dehydrogenase showed that the patient had lowered levels of tissue enzyme as compared to controls. These measurements were made regionally and showed a greater decrease in the brain stem as compared to the cerebral cortex. These changes were shown to be localized, as are the neuropathological changes in this and other cases of Leigh's disease.

"Mixing" experiments were also performed on postmortem tissue. In these studies, liver homogenates from the Leigh's case were mixed in varying amounts with control liver homogenates to determine if there was any inhibitory activity present. Results showed an incremental decrease in activity in the control enzyme activity with addition of Leigh's homogenate. The decrease only reflected the "dilution effect" as would be predicted, not a quick and total depletion expected if there was an inhibitory effect.

Liver homogenates from the Leigh's patient, while showing decreased activity, were able to be activated in vitro by the addition of magnesium chloride and calcium chloride to the homogenate. While the slope of activation was slow, as compared to control values, the fact that it was capable of reactivation suggests that the catalytic components of the pyruvate dehydrogenase complex are present and functional. The "mixing" experiments also argue in favor of this, and argue strongly against the inhibitor concept as pathognomic for Leigh's disease, or even being important.

Results from neuropathological studies showed characteristic lesions as previously described in Leigh's disease patients. Thus, lesions were found in the putamen, caudate nucleus, hypothalamus, substantia nigra, and pontine tegmentum. The mammillary bodies were spared. Microscopically, lesions showed diminished neuropil, neuron loss, demyelination, and an enlargement and increase in numbers of capillaries. In some areas, neurons seemed relatively spared as compared to surrounding neuropil cells. The cerebral cortex was spared, and only minor changes occurred in the spinal cord fasciculus gracilis. These characteristic changes are seen in nearly all Leigh's patients studied.

The authors state that this study suggests dichloroacetate should be considered as a possible treatment. Dichloroacetate has been effective in previous studies in treating lactic acidosis. This study is important because novel biochemical studies were carried out before and after death in an autopsy-proven case of Leigh's disease.

In another interesting report from Japan (Toshima et al., 1982), a case of Leigh's disease, documented at autopsy, was described. The patient was an 11-month-old boy with muscle hypotonia, seizures, and retarded development. Upon admission, the patient had lactic acidosis (elevated threefold), and pyruvate was elevated about the same degree. Also elevated was alpha-ketoglutarate, and alanine was elevated in the urine. Following a glucose tolerance test, lactate/pyruvate levels increased abnormally. Many other enzyme levels were normal, as were CSF values.

A muscle biopsy, taken at 17 months of age showed no evidence of necrosis, or of regeneration. There were no "ragged red" fibers noted. Lipid droplets were seen both with light and electron microscopy. Mitochondria appeared normal. Blood pyruvate dehydrogenase activity was only about 25% that of controls, whereas alpha-ketoglutarate levels were unchanged.

Treatment in this patient consisted of thiamine therapy and therapy with a low carbohydrate–high fat diet. Following 3 weeks of thiamine therapy, there was no improvement in either biochemical or clinical criteria. Therapy with the low carbohydrate–high fat diet for 6 months resulted in a decrease in elevated lactate/pyruvate levels, to nearly control values, and significant clinical improvement. During this trial, weight gain was normal, and the patient's muscle weakness decreased. He became able to sit, and reach for toys. Later, his development quotient dropped, and he died at 25 months from pneumonia.

Results from histopathologic examination at autopsy showed coagulation necrosis and capillary proliferation. The lesions were bilaterally symmetrical and sharply demarcated. Lesions were found in the globus pallidus, caudate, substantia nigra, thalamus, hypothalamus, reticular formation, and cervical spinal cord. Lesions not as striking were present in the cerebral cortex, cerebellum, and putamen. The mammillary bodies were largely unaffected (Figs. 1 and 2).

Nerve cells were involved in some areas, but relatively spared in other areas. Nerve cell changes included pyknosis, karyorrhexis, and eosinophilia. Also present was a striking symmetrical demyelination of nerve cell fibers. The demylination was most pronounced in the reticular formation throughout the brain stem, and cerebellar peduncles. Less demyelinated areas were the internal capsule, cerebral peduncles, and the optic nerve. Also noted was a non-specific meningitis, and it seemed to be related to the lesions of Leigh's disease. This patient was on a respirator for 3 weeks before death, but no lesions of "respirator brain" were found at autopsy.

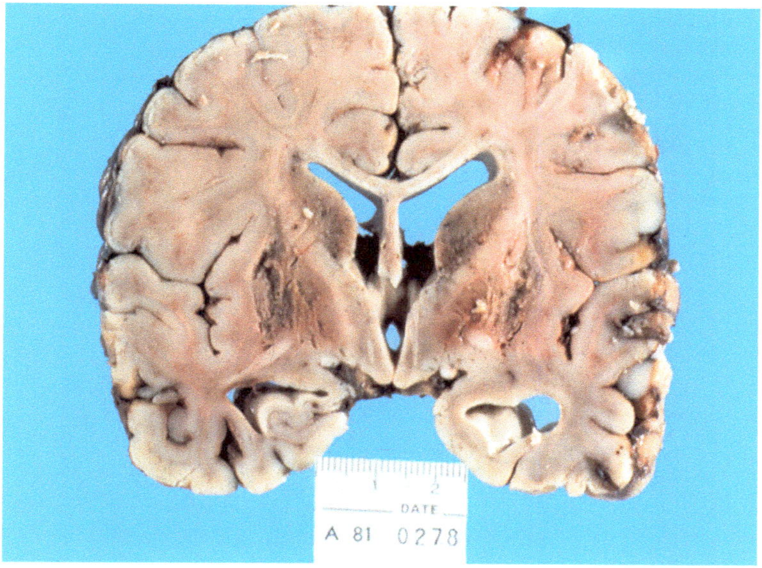

Fig. 1 Photograph of Leigh's brain slice showing lesions in the caudate, putamin, and globus pallidus, but with striking sparing of the mammillary bodies. Figure courtesy of Dr. Michael D. Norenberg

Tissues sampled at autopsy (3 hours after death), were assayed for pyruvate dehydrogenase/decarboxylase activity. Activity measured less than 10% of control values. The enzymes were measured in liver, kidney, cerebral cortex, and the cerebellum.

The idea for treatment with a low carbohydrate–high fat diet is based on the fact that acetyl-CoA can be formed from fatty acids, thereby generating molecules for the TCA cycle which could enter past the diminished pyruvate dehydrogenase complex. Since the patient did show some transient improvement is encouraging, but the cycle did slow down later did occur, with the usual fatal outcome. Similar transient improvement has been observed before. These investigators did demonstrate that the defect in activity of the pyruvate dehydrogenase in brain and liver was similar, whereas activity of pyruvate carboxylase in liver was much higher than that in brain. The authors suggest that a key objective in therapy is to maintain low lactate levels in CSF, indicating an absence of lactic acidosis.

The pyruvate dehydrogenase enzyme complex contains three components and also contains several gene products which may be defective in Leigh's disease (Kerr et al., 1987). The first enzyme component, pyruvate dehydrogenase (E1) is a thiamine-requiring enzyme, and has been found defective in several cases consistent with Leigh's disease.

In the case presented in this paper, a newborn baby was found to have metabolic acidosis, which was corrected. Blood lactate was elevated about threefold. At the time the patient was hypotonic and had diminished auditory response. At the age

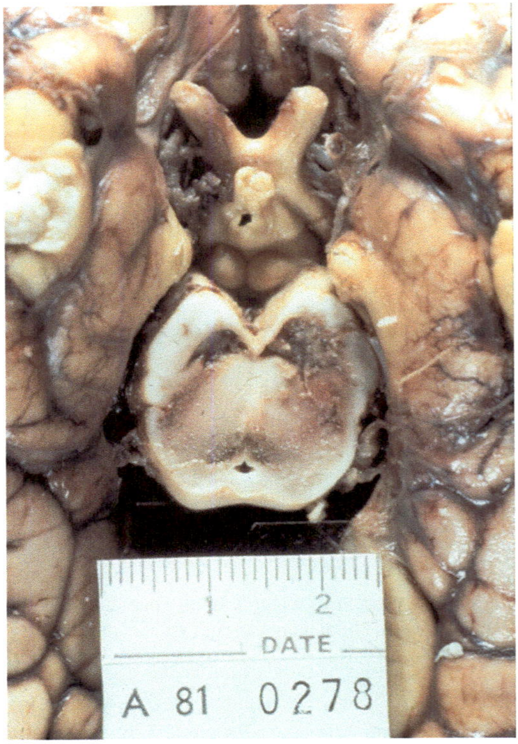

Fig. 2 Photomicrograph of Leigh's brain, lower medulla, fourth ventricle showing symmetric cavitation, and striking vascularity. This slide has a reactive look, and it is symmetrical. Figure courtesy of Dr. Michael D. Norenberg

of 3 months the patient was hospitalized following a brief seizure. At that time, he had strabismus and poor head control. His blood lactate was elevated. Seizures were controlled with phenytoin and phenobarbital, and he had two normal EEGs. Because of the metabolic acidosis, vegetable oil was added to the diet to provide 60% of energy as fat. From 6 to 12 months, he gained weight slowly, and was lethargic. Finally he became markedly lethargic, and had severe metabolic acidosis. His blood lactate levels remained high. He subsequently had a respiratory arrest, was placed on a respirator, and died 2 days later. An autopsy was performed immediately.

Autopsy results showed no evidence of cardiac or skeletal muscle involvement either grossly or microscopically. The CNS showed symmetrical focal areas consisting of cystic necrosis in the putamen and caudate nucleus. The hypothalamus, mammillary bodies, midbrain, and pons appeared normal. Microscopically, neuronal loss and astrocyte proliferation were not seen. There was mild hypomyelination in the cerebral hemispheres and in the optic nerves and posterior columns of the spinal cord. There was also evidence of acute anoxic-ischemic damage. In biochemical studies, PDC deficiency was noted in lymphocytes from blood samples. The degree of deficiency was such that lymphocytes from this patient had activity levels of only about 10% of controls. Tissue samples consisting of liver, heart, kidney, and skeletal muscle were harvested within 1 hour of death, and mitochondria were prepared. These mitochondria showed high respiratory control rates, indicating viability. However, the mitochondria from the four tissue types showed a lack of ability to oxidize pyruvate as compared to controls. In the brain and kidney, whole tissue homogenate's PDC activities were about 1–4% that of controls.

The activity of the three enzymatic components of the PDC complex were measured in fibroblasts, liver, and muscle mitochondria and compared to controls. Results showed activity of E1 in liver was almost zero, and only half in fibroblasts compared to controls. Western blot analysis showed no immunoreactive protein for E1 alpha or E1 beta as compared to controls in samples from liver or brain, and only barely detectable in fibroblasts. By contrast, E2 and E3 components had activity levels comparable to control tissue samples.

The almost complete absence of activity of the alpha/beta subunits of the E1 component of PDC accounts for the overall loss of activity in this patient's tissues. It is the E1 component, which is the rate-limiting step in this complex. The authors note that autopsy results are in keeping with a diagnosis of Leigh's disease. There are some variations in lesions between cases, and in this case, symptoms were present very soon after birth. The rate of onset and rate of progression clearly affects neuropathologic findings. The reason for the reduction of amounts of both alpha and beta components of the E1 complex is not clear. One would expect that a single point mutation would only affect one component. In other published cases, both units have been found to be deficient (Ho et al., 1986), or only the alpha unit was diminished.

Cytochrome c oxidase deficiency is another genetic defect which has been conclusively demonstrated in patients with Leigh's disease (Miranda et al., 1989). This enzyme is a complicated one, with at least 13 subcomponents, some of which are encoded by mitochondrial DNA, others by the nuclear genome. The authors studied fibroblasts from a Leigh's patient who had developed symptoms at the age of 7 months. His parents were clinically normal, but a sibling died with autopsy-confirmed Leigh's disease. Cytochrome c oxidase (COX) activity in fibroblasts was about 18% of normal.

Results showed fibroblasts from the patient after several tissue culture doublings were flattened (instead of spindle shaped), and many contained cytoplasmic vacuoles. SV40 transformed cells from the patient also showed decreased COX activity, similar to untransformed cells. The authors found normal COX activity in hybrids derived from COX-deficient CX3 cells from a Leigh's patient, and COX-positive HeLacot cells, in which the HeLa mtDNA had segregated. The authors interpret this to indicate that the defect in this patient was due to a mutation in nuclear DNA. These results are in general agreement with another study showing no decrease in COX in isolated mitochondria (DiMauro et al., 1987). This study also described "mixing" experiments in which enzyme activity was unchanged when Leigh's mitochondria were mixed with those of a normal control.

Further Leigh's disease research revealed genetic transmission of defects in at least three mitochondrial enzymes, including pyruvate dehydrogenase complex, COX, and, less frequently, NADH Co Q reductase. A T to G point mutation at nt 8993 in the ATPase 6 gene of mtDNA has been described in 12 patients with

Leigh's disease (Santorelli et al., 1993). Symptoms of this defect express as neurogenic weakness ataxia and retinitis pigmentosa (NARP). Santorelli et al. determined that when present in high percentage, the nt 8993 mutation resulted in the Leigh's disease phenotype. The mutation was present in all five mothers studied, but not in maternal grandmothers or aunts. Respiratory chain enzymes were all normal in 11 of 12 patients. The mutation was present in similar amounts in all tissues tested. A distinguishing feature of this mutation is the presence of retinitis pigmentosa. This symptom is not associated at all with COX deficiency or pyruvate dehydrogenase deficiency. In spite of the progressive clinical course of Leigh's disease, the authors suggest that knowledge of this mutation is very important from the standpoint of genetic counseling and possibly for prenatal diagnosis. It is necessary that the phenotype differences between these patients and those with COX and pyruvate dehydrogenase deficiencies are understood.

In a comprehensive paper by Rahmen et al., the specific etiologies of 35 cases of Leigh's disease, and 32 cases with some atypical features, were determined (Rahman, S. et al., 1996). The goal was to determine what proportion of Leigh's patients might have maternal, X-linked, or autosomal recessive inheritance. Muscle samples for analysis were usually taken by open biopsy, or from organ donors (controls).

Results showed that ten patients had DNA mutations at nt 8993 in the mitochondrial genome. Of these, six had a T to G transversion and four had a T to C transition. Enzyme analysis was performed in fibroblasts, and liver and muscle biopsy specimens. Seven patients were found to have pyruvate dehydrogenase deficiency (six had the mutation at the E1 alpha site), nine patients had COX deficiency, and 13 had complex 1 deficiency. No patient had pyruvate carboxylase deficiency. There were some tissue specificity findings in that some pyruvate dehydrogenase deficient patients had only slightly decreased enzyme in fibroblasts, but much lower levels in skeletal muscle, Based on their study, the authors give an estimated incidence of Leigh's disease as 1 in 77,000 births. In an attempt to correlate the clinical picture to the type of defect, no correlation could be made.

At this time (1996), not all Leigh's cases could be assigned to a specific cell/molecular defect, so further studies should reveal other mutations. The present results showed that the complex 1 deficiency was a common cause of Leigh's disease in this study, in contrast to other studies in which it was infrequent. Deficits in several places in the "energy production system" can result in SNE. The actual phenotype in any given patient cannot be predicted by site of mutation; therefore, the onset and severity of symptoms, and the course of the disorder must be related to the degree of impairment of energy producing pathways, and ability of other pathways to compensate. These results, of course, are based on muscle, liver, and body fluids, which may not reflect regional cerebral conditions.

Yet another possible cause of Leigh's disease is mitochondrial DNA depletion (Absalon et al., 2001). In this paper, a patient is presented who was hypotonic at birth and needed resuscitation for several minutes. Physical examination showed poor spontaneous movement of facial muscles, extensor posturing of upper extremities, flaccid lower extremities, and absence of deep reflexes. There was a primary metabolic acidosis. Brain CT scans were normal on days 1 and 7. An MRI on postnatal day 35 showed signals consistent with a diagnosis of SNE. Muscle biopsy was performed, and results showed normal structure with no "ragged-red" fibers. Histochemical stains showed a pale staining of mitochondrial succinic dehydrogenase and COX. A mutation screen of mitochondrial DNA showed no deletions or point mutations. A southern blot showed that the quantity of mitochondrial DNA was less than 5% of normal control mitochondrial DNA. The patient died on day 40, and an autopsy was denied.

The authors comment that mitochondrial DNA depletion has not been carefully considered as a cause of SNE, and only one previous case has been reported (Morris et al., 1996). The molecular basis of mitochondrial DNA depletion is as yet undetermined. Some studies suggest that mitochondrial transcription factor A may be lacking or non-functional, resulting in the mitochondrial depletion syndrome. These authors again stress the importance of defining the exact molecular mechanism responsible for Leigh's disease in each individual case. Genetic counseling is very important to parents of children with Leigh's disease, and this process is facilitated by knowledge of the nature of the exact defect of the disorder. The authors state that mitochondrial DNA depletion is clearly a cause of Leigh's disease and should be considered as a significant cause than has previously been thought.

Another mitochondrial DNA point mutation involving A 8344 G is of interest because phenotypic expression can be quite varied (Berkovic et al., 1991). In a short paper by Tsao (Tsao et al., 2003), a well-documented case of Leigh's disease was associated with the A 8344 G mutation. This patient was born with hypotonia and postnatal developmental delay. He sat at 8 months of age, stood at 12 months, and was able to say five words at 20 months. At that time, he was admitted with asthma and breathing difficulties. He soon showed respiratory failure, lactic acidosis, and progressive mental decline. Over 3 months, his neurological regression led to the loss of all his motor skills. The patient developed Leigh's symptoms, including muscle weakness, external ophthalmoplegia, ptosis, deafness, and a high lactate/pyruvate ratio. He became lethargic and stuporous. Pneumonia led to respiratory arrest and death.

Cerebral MRI at 20 months of age demonstrated symmetric hyperintense T2-weighted signals from the putamen, claustra, thalamus, midbrain, and medulla. These signals were bilaterally symmetrical. The spinal cord was also affected. Magnetic resonance spectroscopy revealed increased lactate in the putamen. Two months later, the lesion size was described as smaller in the putamen, claustra, and medulla, whereas the lesions were larger in the thalamic nuclei, midbrain, and pons. Liver and muscle biopsy tissue showed minimal COX decrease, no ragged red fibers, and enlarged mitochondria. All other enzyme levels were normal, including pyruvate dehydrogenase.

The A 8344 G mutation is usually associated with epilepsy and ragged red muscle fibers. Increased lactate, deafness, and ataxia are also associated with this mutation. The association of Leigh's disease with the A 8344 G mutation represents an atypical phenotypic manifestation. Several other non-typical phenotypes are associated with this point mutation, such as Charcot-Marie-Tooth disease and the Ekbom syndrome. Leigh's disease can result from a seemingly wide variety of genetic mutations, all of which involve the Krebs cycle, or the respiratory chain. This case points out the diversity of causes of Leigh's disease, and stresses again the importance of looking for the cause. Some of this patient's relatives also had the A 8344 G mutation, but had almost no symptoms. This points out the complicating fact that not all individuals with a given point mutation will be symptomatic. How this dichotomy can occur is a complete mystery.

Another paper (Benit et al., 2004) showing a late onset Leigh's disease case reflects the heterogeneity of both the causes and the phenotypes. This was a case of a patient who had a single episode of febrile seizures at the age of 9 months, and then at 9 years, the patient developed a persistent stiff neck. At that time, MRI, EEG, and EMG were normal.

One year later, the MRI showed a high T2 signal intensity in the putamen, white matter, and brain stem. CSF lactate was elevated, and a mitochondrial complex 1 deficiency was found in a muscle biopsy sample. Two years later, the patient developed acute pancreatitis and respiratory insufficiency. The patient expired 18 months later following an acute multisystem failure (Figs. 3 and 4).

Polarographic studies showed low pyruvate plus malate oxidation in skeletal muscle mitochondria from the patient as compared to controls. Activity ratios showed a decrease in complex 1 dependent activities. RT-PCR and D-HPLC analysis of mRNA from skin fibroblasts further characterized the defect. These defects occurred in the NDUFS3 gene subunit involved in the catalysis of electron transfer from NADH to ubiquinone. Of the complex 1 subunits, some defects produce Leigh's symptoms; defects in other subunits can produce cardiac symptoms independent of Leigh's disease. It is possible that only some mutations of subunits produce super oxide over production, which is known to stimulate cardiomyocyte hypertrophy.

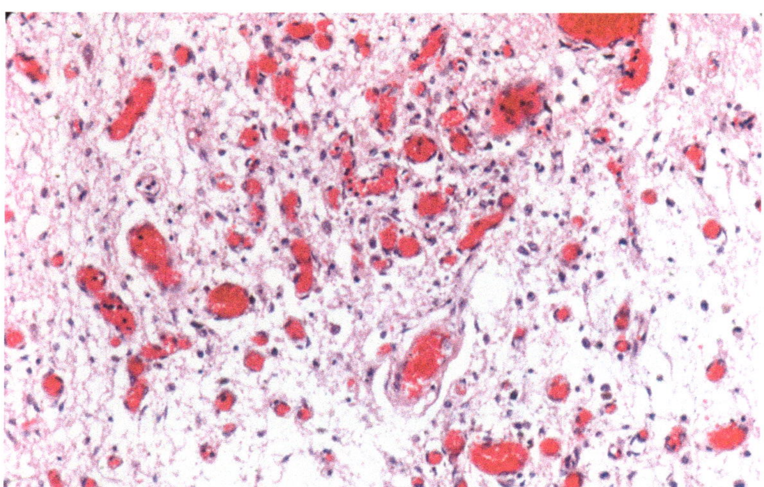

Fig. 3 Photomicrograph of Leigh's brain showing hypervascularity. Figure Courtesy of Dr. Michael D. Norenberg

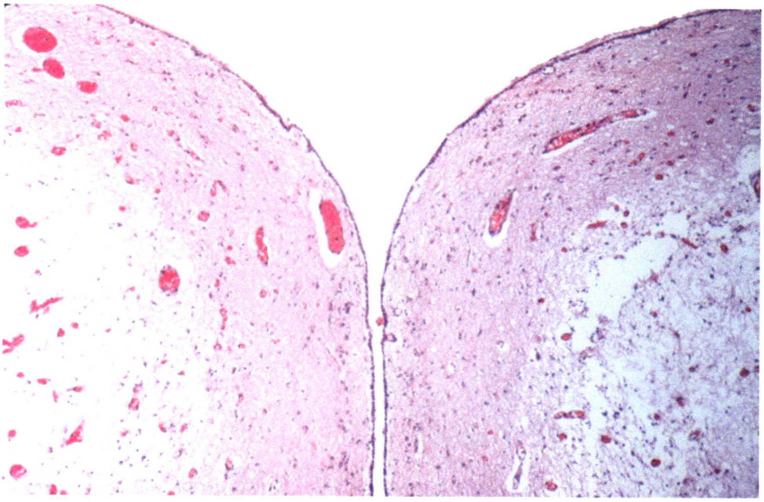

Fig. 4 Photograph of Leigh's brain slice showing brownish lesions in the midbrain. Figure courtesy of Dr. Michael D. Norenberg

Fig. 5 T1-weighted axial image showing swollen hypointense putamina with areas of hypointensity within the lesions. A 1.5-year-old girl with Leigh's disease. Reproduced from Valanne et al. (1998) with permission from the American Society of Neuroradiology

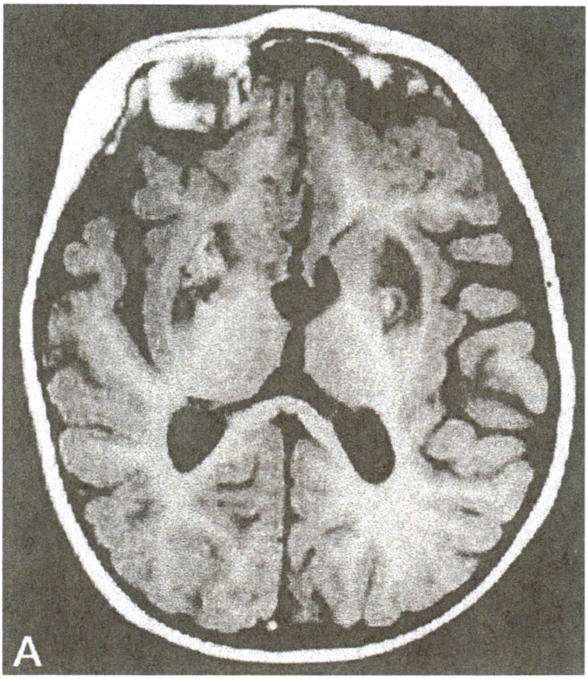

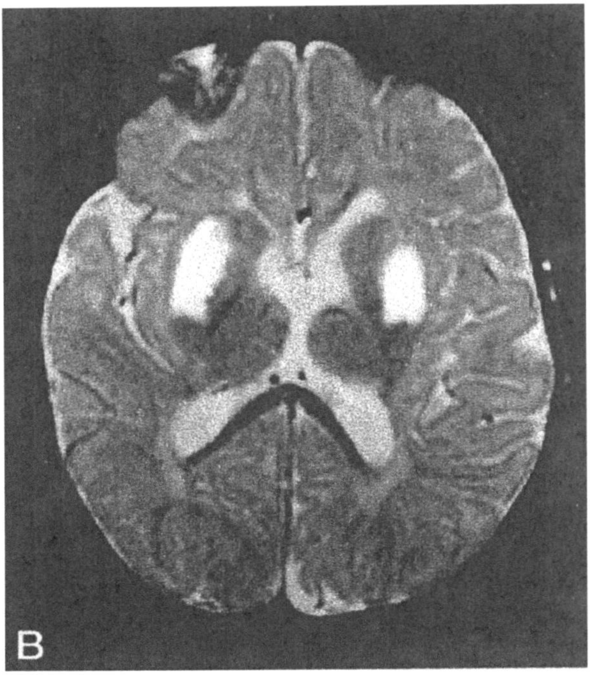

Fig. 6 T2-weighted axial image showing hyperintensity and swelling in the putamina. Myelination delayed 1.5-year-old girl with Leigh's disease. Reproduced from Valanne et al. (1998) with permission from the American Society of Neuroradiology

The above report of a defect in the NDUFS3 subunit represents the last subunit of complex 1 to have a phenotypic reflection of the mutation. All 14 genes (seven mitochondrial, seven nuclear) have disease-causing mutations, several of which may result in Leigh's disease. Other so-called "supernumerary" subunits (at least 28 in number) are associated with complex 1. The functional role of these remaining supernumerary subunits is unclear, but they may be involved in tasks such as regulation of activity of complex 1, protection of complex 1 from oxidative distress, assembly of components, etc. This paper is another example of the very high heterogeneity of causes of Leigh's disease and of the necessity for trying to determine the specific cause of the disorder, at least for counseling purposes.

A recent paper (Hoefs, S. et al., 2008) reports a patient with a mutation in the gene NDUFA2 of the NADH:ubiquinone oxidoreductase (complex 1), the first complex of the respiratory chain. This complex consists of 45 subunits, and it has three functional fractions. The patient in this study was a product of consanguineous parents. An older sibling was normal.

This patient had delayed development from birth and at day 5 had cardiomyopathy. By 4 months, cerebral atrophy and hypoplasia of the corpus callosum showed on examination of MRI images. At 7.5 months the patient developed acidosis, generalized clonic-tonic seizures, and lapsed into a coma. The patient never regained consciousness, and died 43 days later.

Multiple tests were done on this patient. Results showed a defect in mitochondrial energy generating capacity in which there was a decrease in pyruvate oxidation and a decrease in ATP production in the patient's muscle tissue. Immunoblot data showed a decrease in protein level of NDUFA2 in the patient as compared to controls. The mutation in NDUFA2 was associated with the depolarization of mitochondria in living cells. This decreased mitochondrial membrane potential could be reversed by baculoviral complementation. This potential is directly coupled to the normal function of the respiratory chain.

The authors state that complex 1 deficiency is one of the most common defects of the oxidative phosphorylation system. Few treatments are available for these defects, making prenatal diagnosis and genetic counseling very important. The mitochondrial membrane potential could be a hallmark of complex 1 deficiency, making diagnosis more certain.

In another recent study (Sijens, P. et al., 2008), two patients with Leigh's disease were examined with MRI, and lactate was found in gray and white matter, and high choline levels were in white matter only. In the first case, symptoms began to appear at about 9 months. The patient had nystagmus and hypotonia with pyramidal signs. Muscle biopsy showed decreased complex 1, 2, 3, and 4 enzyme activity. This patient died at 3 years of age. The second patient had respiratory problems at 4 years, and examination showed about a 1.5-year developmental delay. Blood lactate was high and he showed a reduced activity of complex 1 and 4, plus diminished muscle ATP production.

Specific results from MRI showed inverted lactate doublet signals were increased in both cases. Choline peaks were high in both patients' white matter. The combination of biochemical data and MRI results led to the Leigh's disease diagnosis in each case. Control patients were those who were not suffering from Leigh's disease, but who had epilepsy, extrapyramidal symptoms, etc.

The findings in these two patients are in keeping with the symptoms and course of Leigh's disease. The authors state that the MRI findings imply altered energy metabolism throughout the entire brain. While choline was elevated in both patients' brains, only one had demonstrable lesions in white matter. While Leigh's disease is thought of as a gray matter disease, these two cases reaffirm the idea that white matter pathology may be occurring. The data suggest that these MRI findings (elevated lactate and choline) could serve as useful diagnostic methods in determining the presence of Leigh's disease.

Leigh's disease and MELAS (mitochondrial encephalopathy, lactic acidosis, and stroke like episodes) have steadily increased in terms of diagnosis and reported cases (Shanske, S. et al., 2008). In these disorders, the G13513A mutation is in the mitochondrial DNA NO5 gene. Results from studies of patients with these disorders showed most but not all had decreased respiratory chain activities at complexes 1, 3, and 4. The ND5 gene is a locale for mutation, and the G13513A mutation was found in 26 of 33 patients. This mutation must be considered in cases of MELAS, or in Leigh's disease. Keys to this mutation include late onset, lack of a family history, and normal or mildly decreased complex 1 activity. It is important to examine multiple sites of maternal tissue in an effort to find the mutation so as to provide good genetic counseling.

African Seasonal Ataxia

Mark 1:6 And John . . . did eat locusts and honey.

Since biblical times man has consumed insects as a primary food source (entomophagy). Insects are, of course, also consumed as a delicacy in many cultures and served in fine restaurants around the world. In Western Africa, and especially in Nigeria, a silkworm larvae, *Anaphe venata*, is consumed in large numbers during the rainy season of July–October. These larvae serve as a primary protein source for hundreds of thousands of indigenous people during this period. Over several years, it became apparent that there were neurological manifestations occurring during the rainy period in those consuming *Anaphe venata* larvae and that this syndrome had many features of thiamine deficiency.

The *Anaphe venata* is a larval stage of the butterfly Lepidoptera: Notodontidae. This larvae serves as a defoliator of *Triplochiton scleroxylon*, which is a forest tree growing in Western Nigeria. During the rainy season these larvae fall from the trees and are easily collected, processed, and sold in the markets (Figs. 1and 2).

Before any correlation of the consumption of *Anaphe venata* and neurological signs and symptoms, a study of the food value of the larvae had been done (Ashiru, 1988). The purpose of the study was to determine the nutritional value of *Anaphe venata* larvae which were being eaten during the rainy season in Nigeria. The larvae are collected and sold in small rural markets.

The investigator in this study had a dual purpose. The first one was to chemically measure the nutrients present in *Anaphe venata* larvae, and the second was to determine the extent of human consumption of the larvae. The food value of *Anaphe venata* was assessed by collecting several hundred larvae and removing the stiff hairs (setae) on the body by passing the larvae through a flame. The larvae were then dried, and powdered in a mortar and pestle and analyzed. Initially the calorific value of the larvae was compared to that of eggs. Results showed that the calorific value fell between that of egg yolk and white. A quantitative nutrient analysis showed that the larvae contained crude protein, fat, ash, manganese, copper, zinc, and phosphorus. When larvae were compared to beef, pork, lamb, and chicken, the larvae had higher protein content than lamb and pork. Six of the eight essential amino acids were found in *Anaphe venata*. Eight other non-essential amino acids

D.W. McCandless, *Thiamine Deficiency and Associated Clinical Disorders*, Contemporary Clinical Neuroscience, DOI 10.1007/978-1-60761-311-4_8,

Fig. 1 *Anaphe venata* larvae generously supplied by Dr. Bola Adamolekun

were identified in the larvae. From these biochemical analyses of nutrients present in *Anaphe venata*, it is clear that they are an excellent source of nutritional value.

In an attempt to determine awareness of the larvae and their use as a food source, questionnaires and interviews were held with people in Nigeria. Results showed that 80% of those interviewed knew the larvae were edible, and nearly 70% had either eaten them or had family members who had consumed the larvae. Before consumption, the setae had been removed by drying the larvae in hot sand. They were then dried in the sun for preservation. When eaten, they were frequently cooked in stew. Entomophagy in the interviewed group had taken place for a range of 9–50 years.

This author concludes that *Anaphe venata* larvae are indeed an excellent source of protein in an otherwise protein-deficient region of Africa. The larvae are only available for a 4-month period, although after drying, they have excellent storage

Fig. 2 *Anaphe venata* larvae generously supplied by Dr. Bola Adamolekun

potential. The author suggests that they would make a useful commercial source for protein, and mass rearing of *Anaphe venata* should be considered.

Descriptions of short-lived epidemics consisting of ataxia were first described in western and north-western sections of Nigeria (Wright, F. and Marley, D., 1958; Coakham, H., 1972; Osuntokun, B., 1972). These brief reports described clinical features of what we now know as African Seasonal Ataxia (ASA), but before the correlation with thiamine deficiency had been discovered. In the report by Osuntokun, it was emphasized that the disorder was not "flu-like," but rather seemed to have a rapid onset a few hours after a meal, usually consisting of yams. In 1971 there were hundreds of cases throughout Western and Northwestern Nigeria. The disorder was described as self limiting, and the first symptoms were nausea and vomiting. The most prominent symptom was a course tremor, with signs of cerebellar involvement. Altered consciousness was present in at least 40% of patients, and ranged from confusion to stupor and coma. Other symptoms included echolalia, myoclonic jerks, abdominal pain, cogwheel rigidity, pin point pupils, etc. There was no evidence of viral infection. The etiology was suspected to be a form of intoxication resulting from a cholinergic stimulation by an unknown substance in the yams. The author speculated that there could be a genetic link in the disorder and that there was no evidence of vitamin deficiencies or intoxication by cyanate or cyanide.

In 1992, Adamolekun was the first to describe ASA in Western Nigeria, and correlate it to thiamine deficiency (Adamolekun, B., 1992). His clinical observations were that the syndrome occurs acutely after patients consumed a carbohydrate meal. Symptoms include dizziness, vomiting, ataxia, and altered levels of consciousness. The affliction occurs in males and females equally, and in all age groups. The impairment of consciousness can end up in overt coma.

Adamolekun states that he was impressed by the similarities between patients presenting with ASA and the symptoms described by Wernicke in 1881 as acute thiamine deficiency (Wernicke, 1881). These classical acute symptoms of thiamine deficiency include ataxia, nystagmus, and alterations of consciousness. Adamolekun noted that the symptoms responded well to Parenterovite 1 and 2, a preparation containing high concentrations of vitamins.

Adamolekun speculates in his first paper on ASA that the patients developing symptoms were marginally thiamine deficient due to the consumption of something possibly containing a thiaminase. The seasonal nature of the affliction suggested the possibility of a thiaminase in the larvae of seasonal insects. A large carbohydrate meal could then trigger the symptoms by placing an unusually high demand for already diminished thiamine reserves.

Soon after the above observations (Adamolekun, 1993), the correlation between ASA and the *Anaphe venata* larvae was made. In those studies, the author interviewed 29 hospitalized patients with a diagnosis of ASA who had developed symptoms during the 1991 and 1992 seasons. All patients were of a low socioeconomic level, and were consuming high-carbohydrate meals, and were protein deficient (mean albumin 32.5 g/L, normal range 35–40 g/L). All patients had consumed roasted *Anaphe venata* larvae as part of their meals immediately before the onset of symptoms. Adamolekun noted that the silkworm *Anaphe venata* moults only once per year and that correlates with the rainy season. The larvae are easily collected, and are sold in the local markets. The author further states that thiaminases are present in many insects and that the interviews of patients suggesting *Anaphe venata* entomophagy and the availability of the larvae as a protein source were highly suggestive.

In 1993 there was an outbreak of ASA, again in Western Nigeria during the rainy season, which was well documented (Adamolekun, B. and Ibikunle, F., 1994). This outbreak took place in a small (population 60,000) town consisting largely of farmers and merchants of low socioeconomic means. The authors verified the diagnosis of ASA in 34 new admissions to five local hospitals. Careful histories were taken and the meal eaten just prior to the onset of illness was noted. Symptomology and a neurological exam were undertaken.

Of the 34 confirmed cases, there were 8 men and 26 women; median age was 29 years. Eight patients were children under 8 years of age. All adults were farmers or small traders of a low socioeconomic class. Presenting symptoms included those previously described, including nausea, vomiting, dizziness, tremor, ataxia, and confusion. Clinical signs included nystagmus, cerebellar ataxia, intention tremors, dysarthria, and impaired consciousness. All patients reported consuming roasted *Anaphe venata* larvae. The registered hospitals in the area reported 1126 admissions for ASA in the 1993 season, a rate of 1.87% in this town.

Patients were treated with IV Parentrovite, a multivitamin solution which contains 250 mg thiamine per dose. Hospital staff were trained in recognizing the disorder and in educating patients and family members of the risks of eating the larvae. It was hoped that the information would be spread among others consuming the larvae, which might decrease the entomophagy.

There was a large increase in ASA during the rainy season in 1993 as compared to previous years. This was attributed to a high inflation rate which made purchase of other protein food sources more difficult. Clinically, all had cerebellar ataxia, and some showed basal ganglia and autonomic dysfunction. There were no deaths attributed to ASA during this reporting period. The incidence rate of 1.87% was deemed falsely low since many afflicted people may have chosen "traditional" non-hospital treatments. As it was, the epidemic accounted for over 70% of all hospital admissions in the region for the month of August.

In another study, a double-blind placebo controlled study was performed examining the efficacy of thiamine as a treatment modality for ASA (Adamolekun, B. et al., 1994). Fourteen patients were assigned randomly to either a placebo group or to a thiamine hydrochloride group. Severity of clinical signs was assessed on admission to the hospital, at the start of therapy, and at every 24 hours for 72 hours. All patients were unable to stand or sit up in bed before therapy started. Patients showing no improvement at 72 hours were treated with multivitamin therapy IV.

Eight patients were treated with thiamine hydrochloride, and six with placebo (lactose). Results showed a dramatic and constant improvement in neurological signs such that in five of the eight thiamine-treated patients, the neurological signs were completely reversed by 72 hours. The remaining three were completely reversed in the immediate post-study period. All patients on placebo deteriorated slightly over the 72-hour period. The improvement in thiamine-treated patients, and the differences between treated and placebo patients was statistically significant.

These results unequivocally implicate thiamine deficiency in the generation of symptoms in ASA. The patients with this disorder consume diets with high levels of carbohydrates which may contain thiamine binding cyanogenic glycosides. Then, during the rainy season when protein-rich *Anaphe venata* larvae fall from the trees, thiaminases present in the larvae are consumed in large amount. This effectively pushes the patients into overt thiamine deficiency producing the ASA epidemics. In fact, a history of monotonous carbohydrate diet and *Anaphe venata* entomophagy was present in all patients examined. The rapid reversal of symptoms with thiamine administration suggest the classical "biochemical lesion" of thiamine deficiency is present before structural changes occur. This rapid reversibility is similar to that seen in early Wernicke's disease when symptoms can be reversed with thiamine treatment. It would be expected that with prolonged thiamine deficiency and/or yearly development of symptoms, structural damage might occur in the brain, as is seen in the Wernicke-Korsakoff syndrome.

The epidemiology of ASA has also been studied (Adamolekun, B. and Ndububa, D., 1994). Results from this study showed that of 37 patients, the ratio of males to females was 1.5–1.0. This ratio could be due to males more likely to come to the hospital, and more likely to consume *Anaphe venata* larvae. The peak month of incidence was August, with some cases appearing in July and September. All patients became symptomatic from 7 to 12 hours after the last main meal. All had consumed heavy carbohydrate meals. All had also consumed a stew containing the *Anaphe venata* larvae. There was no evidence or history of a familial link to the disorder, or of transmission from one family member to another. There was no history of preceding febrile illnesses. The symptoms were the same as described previously, and consisted of cerebellar ataxia, tremor, nystagmus, dysarthria, and impaired consciousness. CSF was normal, but nearly one half of patients showed a diffuse slowing of the EEG. There was no evidence of alcohol or other drug abuse, which might have influenced these patients. These data continued to point to a thiamine deficiency state, and indeed all responded positively to IV vitamin B complex infusions.

In a review article on ASA (Adamolekun, B. et al., 1997), it was noted that the high-carbohydrate diet of ASA patients in Western Nigeria was cassava (*Manihot esculenta*), which contain the cyanogenic glycosides linamarin and lotustralin (Jackson, F. et al., 1992). This same diet is consumed in Zaire, Tanzania, Senegal, Uganda, and Mozambique. Interestingly, several neurological disorders have been associated with ingestion of cassava/cyanogenic glycosides accompanied by depleted intake of sulfur containing amino acids.

At the time of the Adamolekun review article, 100% of ASA patients had consumed the roasted larvae of *Anaphe venata* regularly and in the last meal before onset of symptoms. Entomophagy is a common socially acceptable way for low socio-economic people to supplement their diets throughout Africa. Wars and civil unrest may increase entomophagy in areas where it might not ordinarily be present, i.e., Zimbabwe.

While the *Anaphe venata* larvae have several essential amino acids, it lacks sulfur containing amino acids such as cysteine and methionine. The cyanogenic glycosides consumed in high-carbohydrate foods need sulfur for detoxification. Thiamine, which contains sulfur, could serve as a binding site for the cyanogenic glycosides thereby rendering the vitamin inactive and further contributing to the thiamine-deficient condition.

The exact nature of the mode of "action" of the *Anaphe venata* larvae was speculative until the work of Nishimune et al. (2000). As described in this report, *Anaphe venata* larvae were collected in Uganda. The dry pupae were extracted with sodium potassium buffer by grinding in a motor and pestle. Following centrifugation, the supernatant was used as a crude extract. The extraction was incubated with thiamine for 30 minutes and then assayed for thiamine content. Results showed that thiamine decomposition was time dependent, co-substrate requiring, temperature dependent, and heat labile. Co-substrates were several, including pyridoxine, aniline, and aminopyridine. Based on dialysis experiments, the authors concluded that the thiaminase was a type 1-base exchanging type of thiaminase. The authors found the highest enzyme activity was at 70°C, showing that it was somewhat heat resistant. Heating for 15 minutes at 100°C essentially inactivated the thiaminase. The molecular weight of the thiaminase was estimated to be around 200 kDa.

This study represents the first description of a thiaminase in insects, although these enzymes had previously been described in fish and plants. This study confirmed the hypothesis advanced earlier by Adamolekun that the *Anaphe venata* larvae contained a thiaminase. The fact that the thiaminase in *Anaphe venata* could be largely deactivated at 100°C for 15 minutes suggests that with proper cooking, the larvae could be a safe protein source.

Another study examined the sub-acute effects on behavior of injected extracts of *Anaphe venata* larvae in mice (Iwalewa et al., 2005). Larvae were collected in Western Nigeria, and prepared for consumption in the usual manor. The larvae were crushed and extracted in an aqueous form (polar) and an ethyl acetate form (non-polar). Results from these two extract forms were not significantly different, and can be considered together. The idea was to describe and analyze such acute signs and symptoms such as mild tremor, jerking, hyperactivity, grooming, rearing, sniffing, etc. These signs and symptoms began to show after 7 days of IP injection of the extracts of the *Anaphe venata* larvae.

Results showed that many so-called sub-acute neurotoxic behaviors were not decreased by heat during the processing of the larvae. The internal organs of the mice injected with larvae extracts were not damaged. Observations on organ weights included brain. Results suggested that additional sub-acute toxic effects would produce the full ASA syndrome. This could be expected to occur over a period of months during the rainy season when the consumption of larvae is high.

The observation that mice could be rendered "thiamine" deficient in about 7 days has been noted in a separate study (McCandless, D., 2004). In this study, 25–30 g mice were injected IP with an *Anaphe venata* extract made by pulverizing 1 g of larvae in 2 ml of saline. This mixture was centrifuged and supernatant injected at a dose of 0.1 ml per mouse per day. Mice were also injected with a low dose (50 μg) of pyrithiamine per day. This low dose regime produces neurological symptoms in mice in 10–12 days. In mice injected with both the low dose of pyrithiamine and *Anaphe venata* larval extract, weight loss, ataxia, rolling over movements, and ruffled fur were clearly apparent in 7–8 days. Mice became stuporous, and death occurred by days 9–10. This suggested that a rodent model, similar in symptoms to the human counterpart, was easily available. It is anticipated that future neurochemical studies on brains of animals rendered thiamine deficient by extracts of *Anaphe venata* larvae will show similar results to animals with thiamine deficiency of a more "classic" nature (McCandless, D. and Schenker, S., 1968).

The incidence of *Anaphe venata* induced thiamine deficiency in Nigeria has been reduced primarily due to the efforts of Adamolekun (1992). There was a significant movement to educate the indigenous people as to the risks of *Anaphe venata* entomaphagy. Handouts were prepared, posters placed, and verbal warnings were given by healthcare workers as to the possible risks. These efforts were effective in reducing the frequency of these entomophagy-related outbreaks. They still exist in Nigeria to a lesser extent, and occur in other countries as well.

As an historical note, over 50 years earlier, another thiaminase was discovered in silver foxes among those on the Chastek fox farm in Minnesota (Green, R., 1937). Within a year, the disorder, called Chastek Paralysis, had been experimentally reproduced and the neurological symptoms attributed to thiamine deficiency (3&4 paper, 1942). Characteristically the outbreak occurs during the fall/winter months when the foxes are switched to a diet consisting largely of fish. After a period of 3–6 weeks, affected foxes start to eat less, then progressively develop neurological symptoms consisting of abnormal gait, staggering, and appearances of weakness. Within a day or two, affected foxes develop severe symptoms. These consist of spasticity, severe ataxia, inability to stand, and hyperesthesia. Seizures frequently occur, as well as stupor, and death occurs within about 24 hours following onset of the severe stage.

Pathologically, the heart showed some changes including hemorrhages, which were also seen microscopically. Areas of degeneration were seen in association with hemorrhages. Connective tissue proliferation surrounding vasculature of the heart was also seen microscopically. These cardiac changes were however only seen in about half of the cases. The development of brain symptoms occurred more rapidly, and produced death before the development of most cardiac pathology.

Brain lesions were described as being symmetrical, highly focal, and distributed in a manner reminiscent of Wernicke's disease. In Chastek paralysis, there was proliferation of endothelial cells and alteration of small vessel vasculature. There were many small hemorrhagic areas. The vasculature changes appeared as tortuous vessels which may have reflected the appearance of small vessels which were actually there, but not originally apparent. The changes were strikingly similar to those seen in Wernicke's disease.

In the medulla of foxes suffering Chastek paralysis, lesions were invariably present in the dorsal nucleus of the vagus nerve. Characteristic lesions were seen in the medial vestibular nucleus, and these might extend into other vestibular nuclei. Lesions may also be present in the restiform body and in the floor of the fourth ventricle. Lesions were found in the superior portions of the nucleus gracilis and nuclus cuneatus. Hemorrhages were noted in the reticular formation, but not proliferative vascular changes. The inferior olives were affected in half of the cases examined. The bilateral symmetry was noteworthy, and the degree of change remarkably similar side-to-side.

In the cerebellum, lesions were almost always located in the folia around the forth ventricle. Proliferation and dilation of vessels were prominent in the granular layer, while hemorrhage was more prominent in the molecular layer. The changes in the cerebellum, as in the medulla, were almost always localized in the gray matter and bilaterally symmetric.

In the midbrain, lesions were frequently noted in the inferior colliculus, whereas the superior colliculus was usually spared. The nucleus of the third cranial nerve is usually affected, whereas the fourth cranial nerve nucleus is spared. The red nucleus and substantia nigra were rarely affected. Sections through cortical areas such as the occipital and temporal lobes were negative for lesions. Occasional lesions were seen in the frontal lobes, and these were characteristic vascular changes seen in other areas of the brain. In cortical lesions, as in cerebellar lesions, the most superficial layers were most dramatically affected. Foci of degeneration and of vascular proliferation changes were not seen in white matter.

The authors note the striking similarities between the neuropathology of Chastek paralysis and that of Wernicke's disease. The symptoms of ataxia, nystagmus, stupor, and death are comparable. The authors did not assess eye movement, but lesions support the concept that there would be such changes. As in Wernicke's disease, no lesions were noted that developed outside gray matter, although they may have occasionally extended from gray into white matter. The microscopic changes found in the small blood vessels of Wernicke brains correspond exactly with those seen in Chastek paralysis. These changes in Chastek brains are an irregular dilatation of small blood vessels which give the appearance of vascular proliferation. In both

Chastek paralysis and Wernicke's disease, hemorrhage is noted in the center of the vascular lesion.

The authors speculate about the localization of lesions in Chastek foxes. They state that since thiamine is required for oxidative metabolism in the thiamine-deficient state, pyruvate would be elevated in tissue. Elevated tissue pyruvate levels could in turn be toxic to gray matter precipitating the neuropathology noted. A small number of nursing fox pups also suffered from an acute form of the disorder; however, they had not yet developed structural damage at the time of their demise.

Shortly after the description of Chastek paralysis, the nature of thiamine deactivation in the fish fed to foxes was described (Krampitz, L. and Wooley, D., 1944). In this study, thiamine was incubated with various amounts of carp viscera extract. Following incubation, the amount of thiamine remaining was determined. A variety of paradigms were used, such as varying the pH, using extract of carp viscera as well as whole carp extract, etc.

Results showed that the destruction of thiamine by carp intestine is an enzymatic process. The reaction was largely not affected by either temperature or pH. The enzyme was isolated in 10% NaCl, and is composed of a non-dialyzable, heat labile component, and a dialyzable heat stable component. The thiaminase is therefore presumed to be the component of carp intestine which is responsible for the thiamine deficiency seen in the foxes with Chastek paralysis.

In another thiaminase study, 14 cats were fed a diet of canned commercial cat food which contained whole fish (Jubb, K. et al., 1956). Weight gain in affected cats started to decline after the 11th day on the diet. By the 22nd day, animals had lost 25% of their weight and were barely eating. Ataxia was present a week later and suggested cerebellar involvement. When held by the rear legs, cats' heads were ventroflexed instead of the normal dorsilexed position. Righting reflexes were abnormal, and occasional seizures were seen. Animals, if untreated, quickly become stuporous and die.

Brain lesions were highly focal and bilaterally symmetric. The most severe lesions were found in the inferior colliculus. Gross lesions were also found in the red nucleus, medial vestibular nucleus, habenular nucleus, oculomotor nuclei, cuneate nucleus, mammillary bodies, and other areas.

Microscopically, affected brains showed one or more of three characteristic lesions: varicose dilations of small blood vessels with hemorrhaging, edema, and vascular hypertrophy and gliosis. The vascular dilations were seen not only in lesions but throughout the brain. Lesions were localized in gray matter, and when white matter was involved, it was an extension of the gray matter lesion. Actual hemorrhages were almost always confined to gray matter. In lesions recovering from injury, the thickening of perivascular reticulum and hyperplasia of endothelium gave the appearance of new vessel formation. Edema was confined to the gray matter affected, and rarely seen in white matter. Myelin appeared normal. A glial proliferation was noted in some nuclei which appeared to have been spared any damage by the thiaminase containing cat food. Acute neuronyl changes were not seen within the affected nuclei. Neurons in close association with edema and vascular hemorrhage were usually destroyed.

The authors note that many fish varieties contain a thiaminase which is usually found in the fish viscera. Many different fish parts may be incorporated into cat food, depending on seasonal variations in what is caught. Not all varieties of cat food will contain the thiaminase. The authors speculate that possible focal brain damage might be attributed to accumulating pyruvate due to decreased activity of pyruvate decarboxylase. The striking appearance of the vascular response suggests that it is a very important aspect of the neuropathological lesion. The lesions strongly resemble those seen in both Chastek paralysis and to those of Wernicke's disease. This in itself strongly suggests a thiamine-based syndrome in these cats.

Inherited Ataxias

The inherited ataxias represent a group of disorders which have various etiologies, but have ataxia as a constant and major symptom, and are related to thiamine deficiency. The relation to thiamine deficiency is derived from the finding that many cases have a genetic deficiency in the function of the enzyme pyruvate dehydrogenase. It turns out that there are many variations of inherited ataxias, each with a slightly different biochemical lesion, yet materializing in ataxia.

John Blass was one of the first to describe the inherited ataxias and has been a pioneer in this area. In a case described by Blass, J. et al. (1970), an 8-year-old boy who had several bouts of ataxia per year since the age of 16 months had extensive biochemical tests aimed at elucidating his defect. Clinically, most attacks followed a febrile illness. Symptoms consisted of ataxia lasting several hours to a week. Examinations revealed a cerebellar and choreoathetoid syndrome during acute episodes. In addition, the patient showed irregular eye movements and mild dystonic posturing. These symptoms largely resolved after 4 days.

Pyruvate, alanine, and lactic acid were all elevated in the patient's blood. Alanine was also elevated in urine. Using white blood cells and cultured skin fibrobasts, oxidation of radiolabeled pyruvic acid and pyruvate decarboxylase were all decreased by 80% as compared to controls. Similar studies on the patient's father showed values halfway between controls and the patient's values.

The decrease in pyruvate decarboxylase activity remained low despite addition of thiamine pyrophosphate to the incubation media. This implies that thiamine metabolism in this patient was normal. There was no evidence for the existence of any type of inhibitor of thiamine pyrophosphate. There was an increase in the oxidation of pyruvic acid in white blood cells in both the patient and in controls. This was attributed to an incomplete saturation of apoenzymes and coenzymes.

The authors attribute the symptoms in this patient to the decrease in pyruvate decarboxylase. Elevated pyruvate was the most significant change in the patient's blood. The combination of clinical and biochemical alterations in this patient are uniquely different from those in other similar disorders such as lactic acidosis, Leigh's disease, Hartnup disease, etc. Therefore this patient represented a new description of an ataxia with a biochemical basis.

D.W. McCandless, *Thiamine Deficiency and Associated Clinical Disorders*, Contemporary Clinical Neuroscience, DOI 10.1007/978-1-60761-311-4_9,

A follow-up study was published based on further investigation of this case (Blass, J. et al., 1971). The patient continued having cerebellar ataxic attacks and choreiform movement disorder. Between attacks he was described as mildly clumsy. Otherwise, the patient developed normally, and there was no significant history of head injury, anoxia, toxins, etc., which could have contributed to his condition.

In this particular hospitalization, he was observed to have two ataxic episodes. Symptoms included an ataxic gait, intention tremor, slurred speech, and irregular eye movements. Blood levels of pyruvate were elevated, while lactate levels were usually normal except during some attacks. EEG studies during attacks showed bisynchronous discharges of slow and sharp wave forms without obvious epileptiform activity. Otherwise, EEG patterns were largely normal. Muscle biopsy was normal in paraffin sections; in frozen sections, minor granularity was seen, but no ragged red fibers. Most other lab results were normal.

Two treatment regimes were tried in this patient. First, dexamethasone seemed to have produced a reduction in the duration and severity of ataxic attacks at an earlier age, but not later. Thiamine treatment was associated with a lowering of blood pyruvate. Also, the frequency of attacks decreased from about one per month to a period of 5 months without an attack.

This patient had a clear inherited defect in the first thiamine-dependent enzyme in the pyruvate dehydrogenase (PDH) complex. In other examples of a decrease in the thiamine-dependent component of this enzyme complex, ataxia was present and blood pyruvate levels were increased. While carbohydrates are the major source for energy, ataxia was present, but mentation was normal. Not surprisingly, the defect produced a selective symptom complex.

The authors point out that there seemed to be precipitating events such as fever and stress which push an already marginal enzyme capacity into overt deficiency. The thiamine treatment, which had a measurable positive effect, could be therapeutic because of the incomplete binding between cocarboxylase and the enzyme complex. This binding could be enhanced by saturating the system, thereby increasing the levels of functional complex. The authors point out that a variety of other neurological abnormalities such as Leigh's disease, Wernicke's disease, and Friedrich's ataxia are associated with many of the above symptoms and biochemical abnormalities. Even at this early date (1971), it was suggested that these mitochondrial diseases may have multiple biochemical aberrations.

In another study of a patient with deficient liver pyruvate carboxylase activity, the enzyme was measured in leukocytes and skin fibroblasts and found normal (Brunette, M. et al., 1972). The patient had developed severe seizures 8 hours after birth. Severe hypoglycemia was recognized, and the patient was placed on glucose IV Lactate, pyruvate, and alanine levels were elevated. The patient continued to progress poorly in terms of growth and mentation. Seizures continued, and sporadic acidosis was noted. Thiamine treatment seemed to have an ameliorative effect by decreasing metabolic acidosis. Free and total thiamine levels in blood were normal, suggesting that the beneficial effect might be related to the increased binding of coenzyme to apoenzyme produced by thiamine treatment. It should be remembered that blood levels may not accurately reflect brain tissue levels in discrete regions.

In a review paper (Blass, 1972), the author presents data from over 60 patients and family members showing that parents of patients with pyruvate decarboxylase deficiency have enzyme levels about halfway between controls and patients. The author states that not all hereditary ataxias are related to pyruvate metabolism defects. Conversely, pyruvate metabolism is involved in a significant number of ataxia cases. The author also notes that the properties of PDH appear to be similar between organs, lending credence to using data from fibroblasts to estimate biochemical characteristics of brain enzyme.

Finally, it is noted that biochemical defects of this nature can cause very discrete changes in neural function. In keeping with this is that the significance of biochemical changes is different in various brain regions. The nature of this variation is unclear, but could be related to the presence or absence of reserve capacity of enzyme activity.

Yet another paper (Blass, J. et al., 1972) reports a patient differing from previously reported cases of inherited ataxias in that there was a partial defect in the ability of the patient's fibroblasts to oxidize metabolites through the TCA cycle. This patient was a severely retarded 3-year-old girl. Development after birth was delayed. On admission, the patient had muscular hypotonia, poorly coordinated movements, possible nystagmus, mental retardation, and microcephalis. Blood lactate and pyruvate levels were elevated and thiamine treatment did not lower these metabolites. Citrate levels were also low. Results showed a possible defect in the conversion of citrate to isocitrate, and the activity of PDH was lower in the patient's fibroblasts. Estimation of the first enzymatic component of the PDH complex indicated it was normal. Therefore, the defect must have been in the other two components.

This seems to be a case with severe neurological symptoms, which shows a low conversion to CO_2 of pyruvate, palmitate, and citrate. There was also a defect in the alpha-keoglutarate dehydrogenase enzyme. Although the mechanism of inheritance was not determined, the mother's consanguinity and decreased PDH activity indicates a recessive inheritance. The authors point out that a major oxidative defect could certainly produce major neurological symptoms, as is seen in brief periods of hypoxia/anoxia, or hypoglycemia.

Salient features of a review article (Blass, J. et al., 1975) describing 13 known patients with demonstrated abnormalities of PDH were presented. Most patients presented with a cerebellar ataxia. The movement symptoms were intermittent and frequently precipitated by fever or other stresses. In 10 of the 13 patients, symptoms were clearly identifiable by 1 year of age. More severe neurological symptoms included microcephaly, hypotonia, seizures, and four patients died before the age of 10. Most had elevated levels of blood lactate and pyruvate, or alanine.

In terms of treatment, a high-fat, low-carbohydrate diet seemed to help one patient, but four others became worse. Ketone bodies have the potential to bypass the PDH defect, and feed the TCA cycle. Thiamine treatment at a level of 600 mg per day had little or no effect on the course of inherited ataxias.

Neuropathological studies on the patient who died in 1 year showed many cerebral malformations including an absence of the corpus callosum. Three other patients had diffuse myelin depletion to as little as 10% of controls, and there was a diffuse depletion of neurons. Lesions associated with Leigh's disease were not seen.

Abnormalities in the PDH enzyme complex were identified in most patients. Enzymes such as alpha-ketoglutarate dehydrogenase exhibited normal activities. PDH levels of activity in parents were mostly halfway between controls and patients. Studies of the subcomponents of the PDH complex show that a defect in the first component (thiamine dependent) produces mild ataxia symptoms, whereas defects in the second component produce more severe symptoms. This must relate to the level of impairment of the total activity of PDH. One might expect that a defect in the second component would produce a greater decrease in PDH complex activity. There is also the issue of regional PDH reserve activity in areas such as the cerebellar vermis. In this area it has been shown the reserve activity is lower than in areas such as the cerebral cortex. In keeping with the obvious need for availability of PDH is the finding that its activity in brain mitochondria is three- to fourfold that found in liver mitochondria (Jope, R. and Blass, J., 1975).

Another study by the same group (Jope, R. and Blass, J., 1976) examined the "activity" state of PDH under various conditions. Results showed that ATP concentrations or ATP/ADP ratios influence in an inverse ratio the proportion of PDH in an active form. This was born out examining effects of decapitation, with resultant ischemia. Treatment of mice with amobarbital increased the energy charge and decreased the active form of PDH. These in vivo mouse studies confirm previous in vitro mitochondrial studies (Jope, R. and Blass, J., 1975). The authors note that other factors modulate the activity of PDH, such as the concentrations of Mg and Ca.

One other comment is that the activity of PDH is low relative to the flux of pyruvate. The PDH activities are around 100–225 micromoles/gram tissue/hour (McCandless, D. and Schenker, S., 1968; Jope, R. and Blass, J., 1975). Flux of pyruvate is about 70–140 micromoles/gram tissue/hour. This provides an approximate threefold reserve, in most brain areas, but it should be remembered that some areas have a much lower reserve capacity (Reynolds, S. and Blass, J., 1976).

There may also be a relation between PDH activity and acetylcholine synthesis (Gibson, G. et al., 1975). When various inhibitors such as bromopyruvate were added to minced brain, a decrease of pyruvate oxidation by 5% led to a decrease in acetylcholine by 7%. This also held true for the inhibitor 2-oxobutyric acid (Blass, J. and Lewis, C., 1973).

In a paper from France (Saudubray, J. et al., 1976), two sibs are described with neonatal lactic acidosis associated with a defect in pyruvate carboxylase. Both sibs showed symptoms within 24 hours of birth. These included hypothermia, hypotonia, anorexia, and metabolic acidosis. Blood values of ammonia and lactate were elevated. Pyruvate was mildly elevated, and the lactate/pyruvate ratio was elevated. Several enzymes were decreased; most significantly, pyruvate carboxylase was decreased to as low as 2% of controls.

Although autopsy was refused on the first sibling, permission was granted for the second. Results showed a widespread diffuse and symmetrical demyelination of both cerebral and cerebellar white matter. Symmetrical paraventricular cavities were found associated with the lateral ventricle and also between the dentate nuclei and the fourth ventricle. If there was any other neuropathological data, it was not published. The authors state that these brain findings are not the same as is seen in Leigh's disease.

Clinically and biochemically, the authors suggest that their two cases of pyruvate carboxylase deficiency are unique, and have a rapid clinical course, severe metabolic acidosis, and hyperlactic acidemia. These are all different from Leigh's disease and so fall among the many variations of inherited ataxias.

Also under the category of inherited ataxias are a group characterized by spinocerebellar degeneration, called Friedereich's ataxia. This disorder may result from several mutations, but a significant number involve the oxidation of pyruvate (Blass, J. et al., 1976). And if to emphasize the diversity of symptoms and neuropathological findings, at least one case of clinical Friedreich's ataxia has been published in which the neuropathological findings were those of Leigh's disease (Dunn, H. and Dolman, C., 1969).

In the description of five cases of Friedreich's ataxia, all had ataxia develop around the age of 12. Most patients showed speech slowing. Muscle strength was decreased, and deep reflexes were decreased. Visual findings included nystagmus, diplopia, and decreased corneal reflexes.

Results of biochemical studies showed that the activities of pyruvate and alpha-ketoglutarate dehydrogenases were decreased in disrupted fibroblasts of the 5 patients as compared to 16 control subjects. PDH activity was 43% of controls, while alpha-ketoglutarate dehydrogenase was 50% of control values. Increasing substrate or cofactors in vitro did not change the activity rate. These above data seem to define a biochemical lesion in cases with clear clinical evidence of Friedreich's ataxia.

The authors speculate that the one common factor in both the above enzymes is lipoamide dehydrogenase. A defect in this third component of the enzyme complexes could therefore affect both PDH and alpha-ketoglutarate dehydrogenase, in turn producing the clinical picture of Friedreich's ataxia. Since the brain relies on oxidative metabolism for normal function, decreasing activity of these two critical enzymes associated with the TCA cycle could easily be seen as deleterious to normal function. Many previous studies support the concept that mild–moderate decreases in TCA cycle enzymes are associated with neurological symptoms such as ataxia (Figs. 1–7 are representations of slices and sections from a case of Friederich's ataxia. Courtesy of Dr. Roland Auer).

An important paper looking at methods for measuring activity of PDH in fibroblasts has been published (Sheu, K. et al., 1981). This study examined activities of PDH in normal fibroblasts and in two cell lines originally described by Blass, from patients deficient in the enzyme PDH. It was shown that activation was possible by pretreating fibroblasts with dichloroacetate. The concept is advanced that a wide variety of metabolic situations may have effects on the activity of PDH. Pre-activation of enzyme activity in fibroblasts with dichloroacetate maximizes the enzyme, so that low levels on the order of 3% of normal can be measured. Results

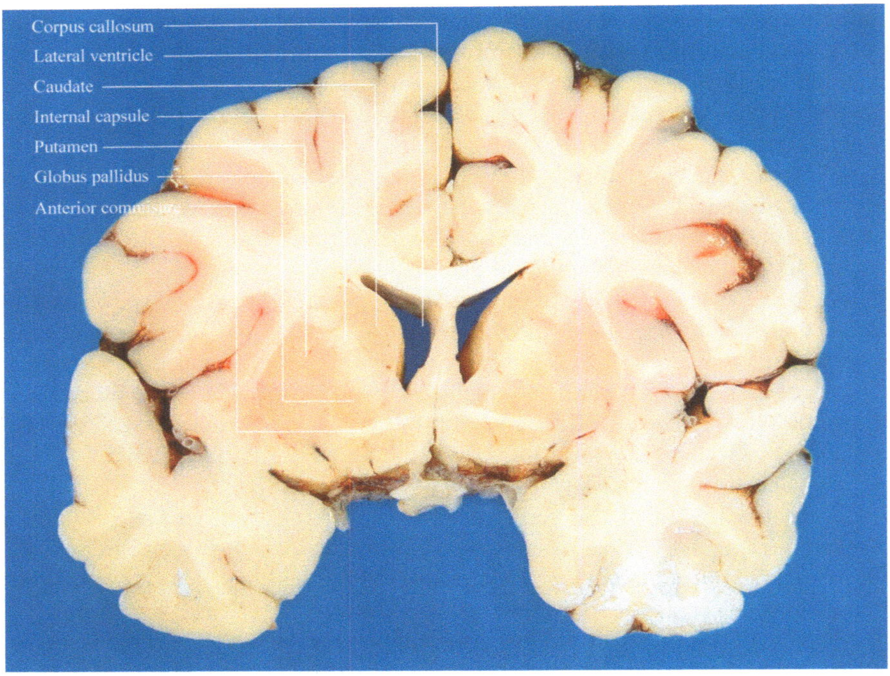

Fig. 1 Brain Slice at the level of the caudate nucleus. Courtesy of Dr. Roland Auer

have confirmed that PDH exists largely in an inactive state and needs activation in order to be measured. The authors point out that in Friedreich's ataxia, treatment of fibroblasts aids detection of the deficiency.

The dichloroacetate activation procedure provides a simple reproducible method to measure PDH activities in inherited ataxia, parents of patients, and normal controls. This should help identify patients at an early stage. These assay techniques may not be available at all hospitals, and any improvement in assay methodology is welcome. The authors are working on techniques to use cells such as leukocytes and amniotic cells.

Central nervous system dysfunction has been described in a case showing congenital lactic acidosis, associated with a defect in complex 1 of the respiratory chain (Moreadith, R. et al., 1984). In this case, a defect in complex 1 on the inner mitochondrial membrane was documented in liver, kidney, heart, and skeletal muscle. The patient was hypoglycemic, and in some respiratory distress within 24 hours of birth. He had cardiac hypertrophy, and liver enlargement by 6 weeks. Neurologically, the patient was unresponsive and no longer reacting to visual or auditory stimuli. A computerized tomographic scan and sonogram showed decreased density of frontal and parietal white matter. There was a concomitant lactic acidosis and an elevated lactate/pyruvate ratio. Despite several therapeutic attempts, including thiamine, the patient expired at 16 weeks of age.

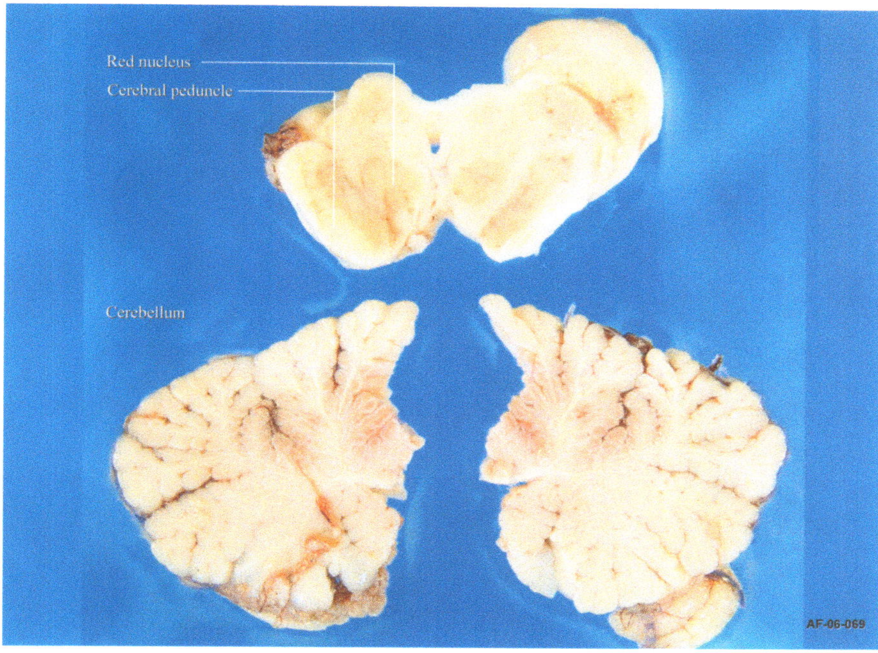

Fig. 2 Friedreich's ataxia case showing cerebellum. Figure courtesy of Dr. Roland Auer

The decreased activity of complex 1 in the mitochondrial respiratory chain was confirmed in liver, heart, kidney, and skeletal muscle. The presence of central nervous system dysfunction and tomographic evidence certainly suggest brain involvement. Measurements of other potentially defective enzymes such as PDH showed normal levels of activity. Tissues examined had fat droplets, probably representing an inability to oxidize fatty acids. The authors state that ATP synthesis by oxidative phosphorylation would be decreased and that depleted ATP would have contributed to the demise of the patient. The authors state that autopsy results suggest brain involvement, although scant neuropathological data are presented.

In another case of PDH deficiency, all three components of the enzyme complex were measured and the defect localized to the first (thiamine requiring) component (Ho, L. et al., 1986). In this study an immunoblot technique was used to assess components of the PDH complex. E1 (the first component) had an activity level of only 33% of control values.

In the over 80 cases described at this point, as many as 50 have been attributed to defective PDH components. The previous measurements of the E1 complex were difficult due to its contributing only about 10% to the overall total PDH activity. The immunoblot method described should simplify and increase the number of defective components of the mutation of PDH.

Emphasizing the complexity and continuing evolving discovery of additional mutations involving inherited ataxia, elevated lactate/pyruvate ratios, and in these

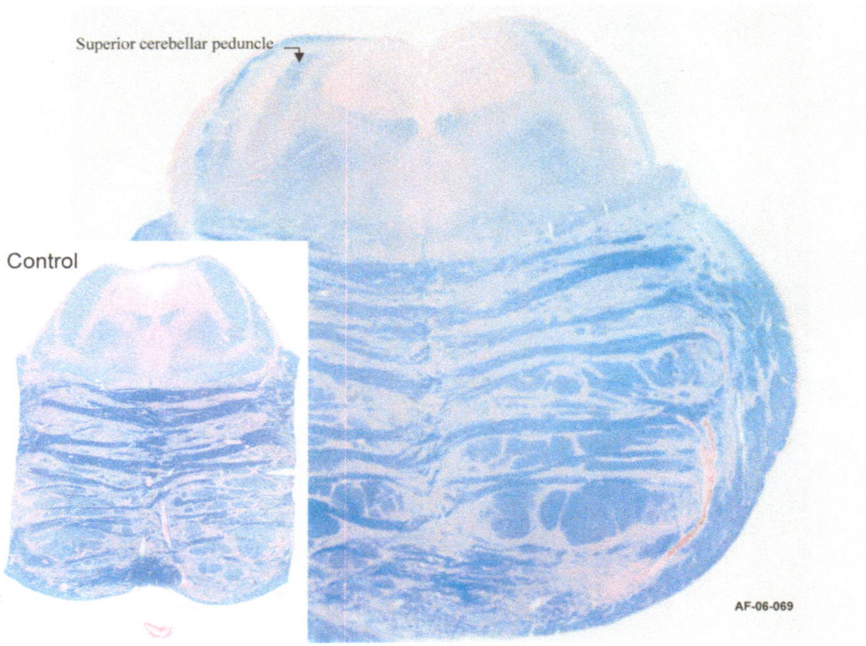

Fig. 3 Brain slice at the level of the superior cerebellar peduncle from a Friedreich's ataxia patient. Figure courtesy of Dr. Roland Auer

cases, seizures may occur (Mollet, J. et al., 2008). Symptoms in these cases were similar. Patients seemed to progress well for a period, but in one case became ataxic by the age of 2. At the age of 2.5, seizures (generalized tonic) began and were accompanied with increased CSF lactate. Sodium valproate and rivotril controlled seizures for 5 years. His neurological condition deteriorated by 12 years of age, such that he could not walk or speak. Seizures were increasing, and MRI revealed a severe cerebellar atrophy along with cerebellar ataxia and ptosis. A brain MRI showed cerebellar vermis hypoplasia. Some years later, he showed severe cerebellar atrophy. Yet another patient in this study is a sister of case 2 and had a similar course. At the age of 13, she had cerebellar ataxia.

Numerous molecular studies described in this manuscript showed a defect in the CABC1 gene which is involved in the ubiquinone biosynthesis pathway. The precise function of the ABC1 protein is unclear. The patients in this study were all ubiquinone deficient. The deficiency was noted in skeletal muscle; no "ragged red" fibers were noted.

A key clinical observation in those patients was the development of seizures, to the point of "epilepsia partialis continua." The authors looked for the CABC1 mutation in patients with similar symptoms (such as Leigh's patients) and the CABC1 mutation was not found. This indicates a new variation of cerebellar ataxia attributed to the CABC1 gene, and not previously described. Although rather similar in clin-

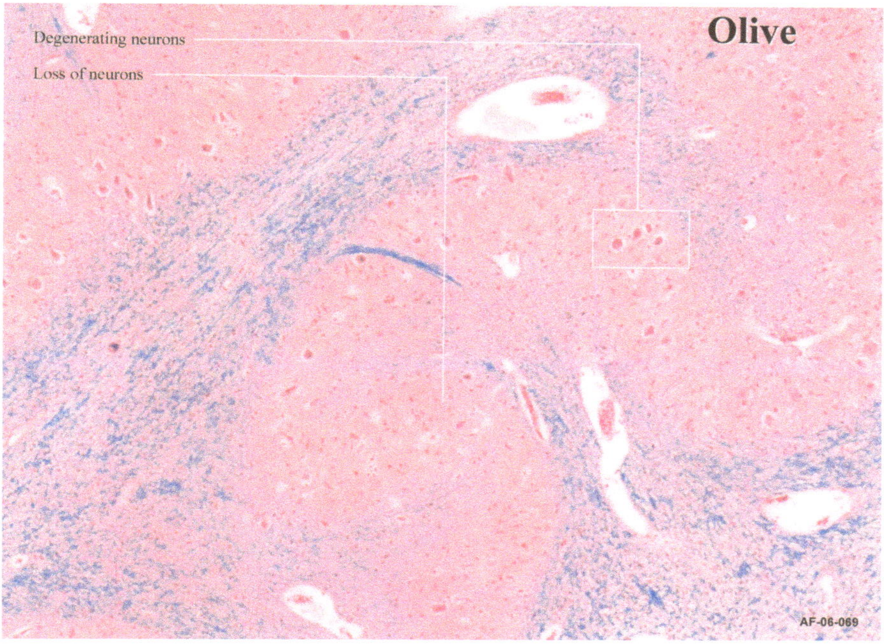

Fig. 4 Photomicrograph of the olive in a Friedreich's ataxia patient. Figure courtesy of Dr. Roland Auer

ical presentation, the patients had varying degrees of respiratory chain dysfunction and ubiquinone deficiency. For example, two patients had severely decreased muscle ubiquinone content, whereas another patient had ubiquinone levels at lower control values. The authors state that muscle values may not reflect metabolism in the CNS. The exact function of the CABC1 protein is unclear, especially in brain. Ubiquinone treatment trials had little beneficial effect.

The reason for variations in symptoms and in tissue responses to mutation of the CABC1 gene is unclear. Probably different organs would have different ubiquinone requirements and have different normal ability to metabolize ubiquinone. These tissue differences would of course be reflected in the deficient state. Additional hypotheses to explain different tissue responses involve expression levels of various proteins concerned with ubiquinone synthesis and metabolism. None of the patients with CABC1, SCo2, or p53R2 mutations showed any evidence of apoptosis or tumor susceptibility. CABC1 is a target of the p53 tumor suppressor gene, making it the third p53 gene in disorders of mitochondria (Iiizumi, M. et al., 2002).

To summarize, this manuscript reports the first description of severe neurological symptoms, and cerebellar atrophy/ataxia attributed to a point mutation. Treatment regimens were unsuccessful. Severe seizures to the point of epilepsia partialis continua were a unique feature of seizures in these cases. The seizures seemed intractable to drug therapy, but other seizure treatment regimes were apparently not tried. Again, these cases point out the extremely complex features of these inher-

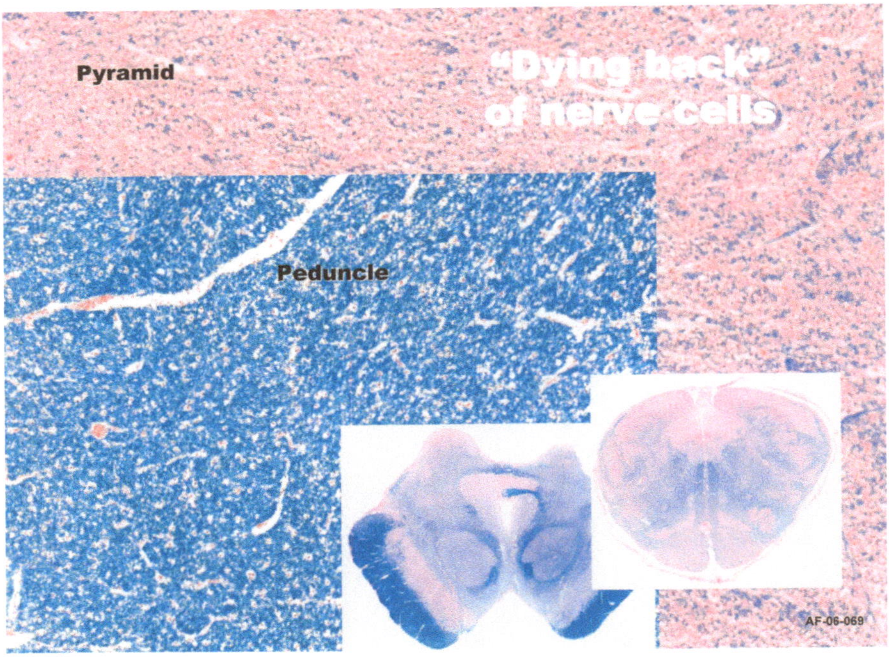

Fig. 5 Photomicrograph showing "dying back of neurons" in a Friedreich's ataxia case. Figure courtesy of Dr. Roland Auer

ited ataxias and the problems of both diagnosis and possible treatment modalities. Correct diagnosis alone and delineation of the disorder biochemical features outside of a major medical center seem daunting.

A recent review paper (Sedel, F. et al., 2007) has been published looking at inborn errors of metabolism. Examined were a large number of inborn errors of metabolism (IEMs) with emphasis on those amenable to treatment. These were characterized and assigned to various groups including inherited ataxias. It is stated that nearly all IEMs affect many organs such as liver, heart, kidney, skeletal muscle, but nearly all also affect brain. IEMs are candidates for various treatment modalities, including enzyme and gene replacement therapy. The review focuses on disorders that are mild–moderate in symptom presentation and have late childhood or adolescent onset.

Inborn errors of metabolism which involve chronic ataxias include the PDH deficiency syndromes. In these, MRI of the CNS usually shows bilateral symmetric lesions in some but not all ways similar to lesions seen in Leigh's disease. These disorders characteristically have cerebellar ataxia, ocular disturbances, deafness, lactic acidosis, and elevated lactate/pyruvate ratios. Successful treatment of these specific IEMs have been elusive. Often, acute episodes of cerebellar ataxia can be triggered by stress or fever in PDH deficiency patients. Chronic cerebellar ataxia also accompanies Friedreich's ataxia and coenzyme Q10 deficiency. Seizures may

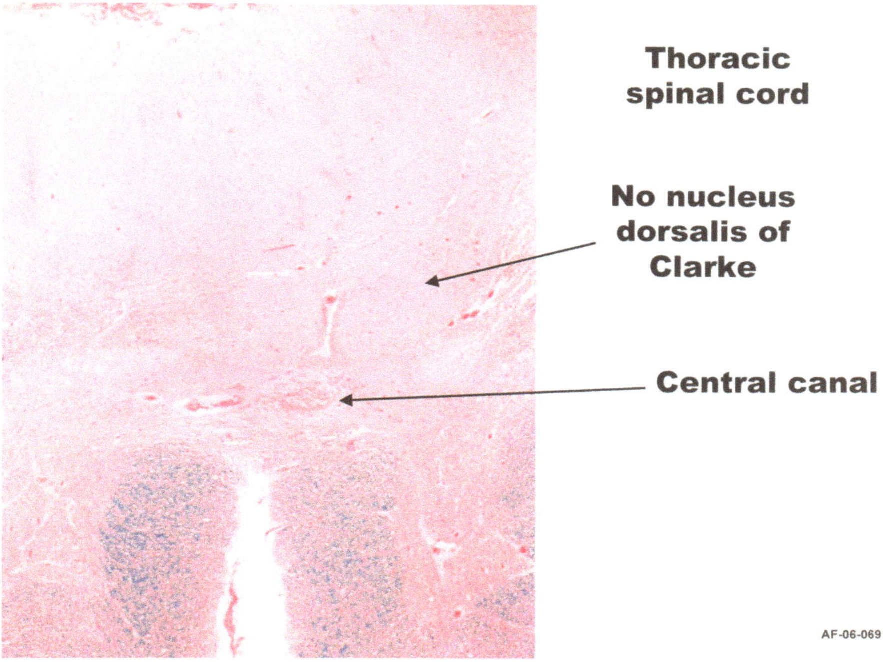

Fig. 6 Slice at the level of the thoracic spinal cord from a case of Friedreich's ataxia. Figure courtesy of Dr. Roland Auer

occasionally accompany inherited ataxias and may respond to typical anticonvulsant therapy. Seizures are characteristic in glucose transporter (GLUT-1) deficiency. In the case of GLUT-1 deficiency seizures, treatment with a ketogenic diet may be effective. These patients may also show loss of muscle tone and impaired consciousness. There is usually a low CSF/blood glucose ratio, which can be useful in diagnosing this IEM.

In a review paper (Berendzen, K. et al., 2006) the use of oral dichloroacetate in patients with congenital lactic acidosis was examined. Much of the lactic acidosis is a result of improved pyruvate dehydrogenase activity, and about 90% of cases are due to defects in the thiamine requiring first component of the three-component haloenzyme. At least 80 different mutations have been reported in patients with lactic acidosis. Inhibition of this key enzymatic step is thought to decrease energy production in brain.

Dichloroacetate increases pyruvate dehydrogenase activity by either inhibiting pyruvate dehydrogenase kinase and/or by stabilizing the dehydrogenase enzyme or slowing its rate of breakdown. Looking retrospectively at patients' records it was shown that those receiving a daily dose range of 20–135 mg/kg of dichloroacetate was associated with a decrease in blood and CSF lactate by about 40%. Four patients

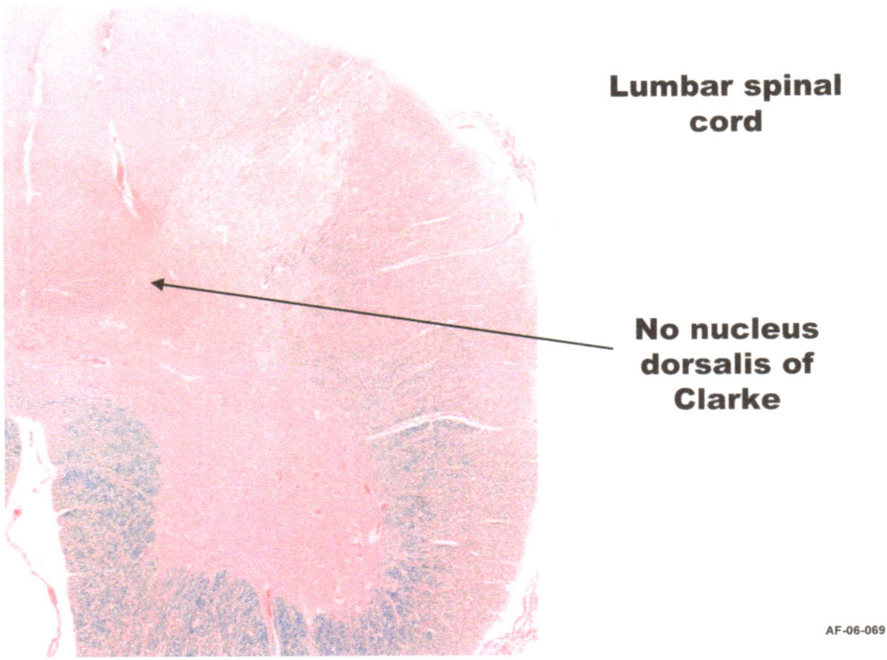

Fig. 7 Slice at the level of the lumbar spinal cord from a case of Friedreich's ataxia. Figure courtesy of Dr. Roland Auer

with pyruvate dehydrogenase complex deficiency have been treated for up to 5 years and are still being evaluated.

The authors argue that this data supports the concept that long-term (5 years) treatment with dichloroacetate is safe. The exact benefit for these disorders is unclear, but controlled clinical studies seem warranted based on the above data on dichloroacetate.

In a paper (Debray, F. et al., 2008), two siblings with ataxia were studied and found to have an unreported mutation (G585C) in the pyruvate dehydrogenase A1 gene. These patients clinically exhibited intermittent ataxia. The clinical heterogeneity of pyruvate dehydrogenase deficiency is emphasized, and these two patients may represent the first two cases in which CSF lactate levels were not elevated. This disorder seems to have a broad spectrum of clinical progression. In the mildest form there may be occasional episodes of ataxia, with complete resolution between episodes. In the worst case, the patient has a course as seen in Leigh's disease, relentless and resulting in death.

This paper describes a heretofore unreported defect in the pyruvate dehydrogenase A1 gene where the mutation c6585c leads to the replacement of glycine to alanine. This change is in a thiamine pyrophosphate binding domain (TPP-BD). Thiamine treatment had no improving effect on the two patients. The authors emphasize the large phenotypic variation in symptom presentation in intermittent ataxias and suggest that the intermittent nature of these disorders could lead to an underestimation of incidence.

In an interesting recent case description (Henwood, M. et al., 2006), a patient previously diagnosed and well documented with pyruvate dehydrogenase deficiency presented with severe diabetic ketoacidosis. She was being treated with a ketogenic diet aimed at circumventing in part the impaired dehydrogenation of pyruvate with resultant energy metabolism depletion. The dilemma was to find treatments for the pyruvate dehydrogenase deficiency and diabetes.

The authors implemented a ketogenic diet and limited insulin therapy which achieved a safe balance. It was shown that adequate ketosis and insulin treatment could coexist such that each metabolic disturbance could be satisfactorily treated. In fact, her progress under this new treatment regime has improved overall including her mental status. In addition, her growth has improved, as well as developmental milestones, such as sitting up, standing, and drinking from a cup.

In a recent study (Kang, H. et al., 2007) the characteristics of seizures were examined in patients with mitochondrial diseases. This was a retrospective study in which 142 patients were examined in terms of seizure phenotype and mitochondrial disorder. Of these patients, 45 were identified as having mitochondrial disease, and 22 of these were epileptic. Of the 22, various clinical manifestations were present including optic neuropathy, hearing loss, cardiomyopathy, developmental delay, etc. Eight patients in whom the test was done showed significant decrease in complex 1 activity. The epileptic phenotypes were diverse. Some had severe myoclonic seizures, some generalized seizures, some had absence seizures, etc. Seizures, however, were the first unequivocal symptom in 16 patients. All of these patients had EEGs characteristic for the phenotypic type seizure. These findings are not unusual in that it is recognized that mitochondrial diseases can lead to brain damage conducive to seizures (Kunz, W., 2002). In some cases the diagnosis of mitochondrial disorders precedes seizures, and of course not all patients with mitochondrial disorders ever have seizures.

Most of the patients described in this manuscript did not respond well to antiepileptic drug treatment. The treatment of mitochondrial disorder patients with ketogenic diets has been successful in some patients, but not all. The authors point out that the retrospective data cannot provide definitive numbers concerning the number of epilepsy cases in mitochondrial disorders. Data does suggest that mitochondrial disorders were more common in epileptics than anticipated. While the phenotype expression of these epileptic patients is diverse, the clinician should always be alert to the possible coexistence of both disorders in the same patient.

In a retrospective study, 73 patients diagnosed with mitochondrial diseases were studied (Debray, F. et al., 2007). Results showed that the patients could be arranged in seven phenotypes: (1) neonatal onset lactic acidosis (10%), (2) Leigh's syndrome (18%), (3) intermittent neurologic (5%), (4) Leber hereditary optic neuropathy (5%), (5) mitochondrial myopathy (19%), (6) visceral (11%), and (7) nonspecific encephalopathy (32%).

Initial clinical presentation occurred at a mean age of 7 months and included metabolic acidosis, failure to thrive, seizures, ataxia, muscle weakness, psychomotor delay, muscle weakness, etc. In the neonatal group, all presented shortly after birth with lactic acidosis and neurological distress. All died in the neonatal group with a median age at death of 4 days.

Biochemical studies showed that 72% of the patients had chronic lactic acidosis. Blood alanine was elevated in a similar number of patients. Krebs cycle intermediates were elevated and ketone bodies were found in the urine of 70% of patients in which these were measured.

MR images were available in 59 patients, and most frequently seen were basal ganglia hyperintensities and cerebral atrophy. Other significant findings included those localized to the cerebellum, brainstem, white matter, and corpus callosum. These were all hyperintensities. Stroke like infarcts were occasionally seen. Muscle samples examined with electron microscopy showed mitochondrial proliferation or ultrastructural morphologic alterations in 80% of 40 patients' samples. Staining techniques showed mitochondrial proliferation in about 50% of samples examined (Figs. 8, 9, 10, and 11).

Biochemically, an enzyme defect was identified in 44 patients, and complex 4 deficiency was the most frequent. Pyruvate dehydrogenase and complex 4 were the most common defects, about 25% each. The overall mortality rate was 46% with a median survival rate of 12 years. There was no association of mortality rate and the

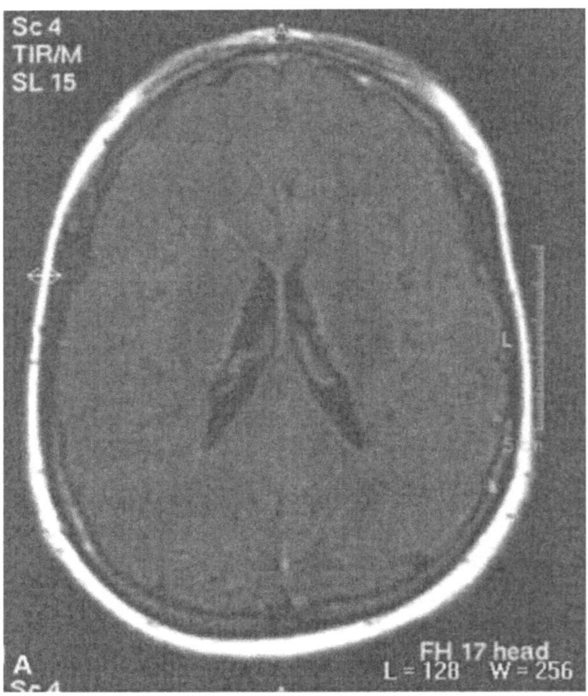

Fig. 8 MRI of a Leigh's patient showing high signal intensity of periventricular white matter. Reproduced from Schiff et al. (2006) with permission from Wiley-Blackwell

Fig. 9 MRI of a Leigh's patient showing high signal intensity from swollen caudate nuclei and putamina and thalamus at acute deterioration. Reproduced from Schiff et al. (2006) with permission from Wiley-Blackwell

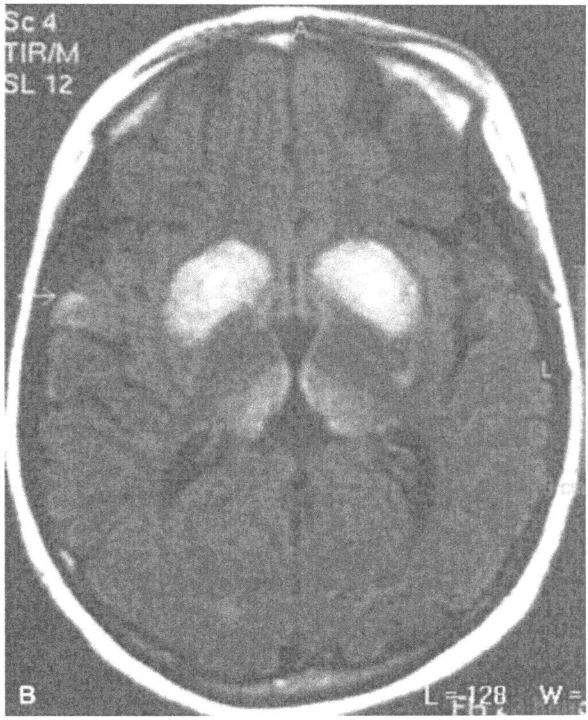

Fig. 10 MRI of Leigh's patient showing high signal intensity of cortex and subcortex at a time of acute deterioration. Reproduced from Schiff et al. (2006) with permission from Wiley-Blackwell

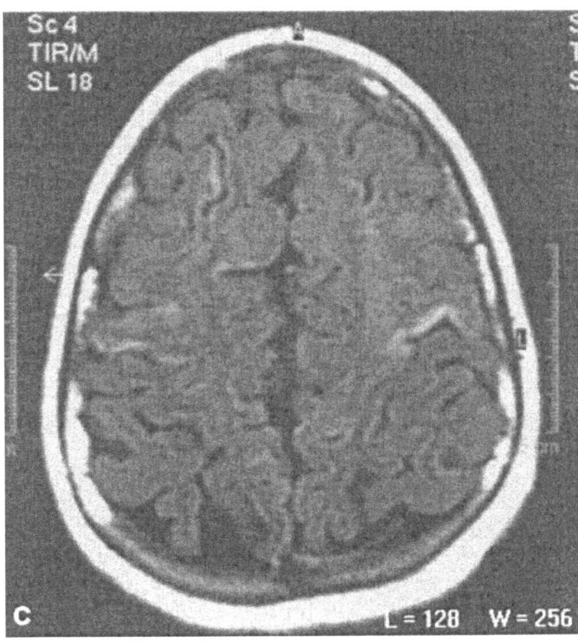

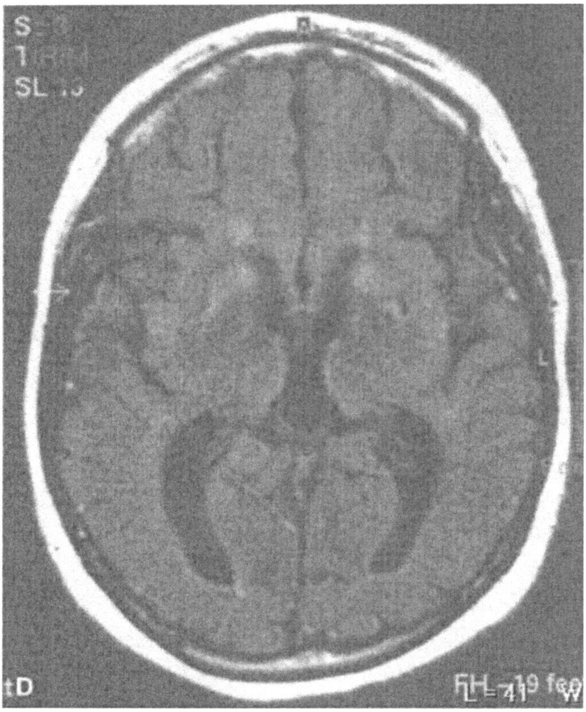

Fig. 11 MRI of a Leigh's patient 1 year after acute deterioration showing bilateral hypersignal and atrophy of the basal ganglia, and enlarged ventricles with cerebral atrophy. Reproduced from Schiff et al. (2006) with permission from Wiley-Blackwell

biochemical defect. Signs before 6 months of age of liver involvement, metabolic crisis, etc., were indicators of high mortality rate.

The authors point out a high frequency in these patients of a complication of some sort in the pregnancies. A low birth weight was also noted. Cardiomyopathies were relatively low in these patients. Measurement of blood lactate proved to be a non-specific test. Since most of these disorders are untreatable, the authors suggest initially using the least invasive techniques, which can have a good yield in terms of diagnosis. The clinical course of these disorders can be long term, so clinical follow-up is essential. The authors state that many patients have only mild or no sequelae after 5 years of follow-up. This paper again points out the "overlap" between Leigh's disease and the so-called inherited ataxias.

In a recent case (Sedel, F. et al., 2008), an adult with a deficiency of pyruvate dehydrogenase was successfully treated with thiamine. The patient was a 26-year-old man who from about the age of 2 had paroxysmal walking problems associated with fever. He had optic neuropathy by the age of 20, and measurement of blood lactate showed a mild increase. Measurement of lymphocytic pyruvate dehydrogenase activity showed a level of only 13% of control. Sequencing the PDHA1 gene showed a mutation in the third exon (E75A). This is a mutation not recorded before. No other mutation in the PDHA1 gene was found.

Treatment with thiamine at a dose of 1 gm/day was initiated. Eighteen months later, visual acuity was improved, and no additional episodes of muscle weakness or ataxia had occurred. Lactate levels were within normal limits. The effects of treatment on ataxia and muscle weakness are thought by the authors to be a direct effect of thiamine treatment. Adding thiamine to the patient's fibroblasts in culture increased pyruvate dehydrogenase activity nearly fivefold. The observations are significant because pyruvate dehydrogenase activity has only rarely been described in adults, and especially because the deficiency has only rarely been improved with thiamine treatment. It should be noted that the dose of 1 gm/day is a very high dose. These positive findings are unique in that this mutation may be equally unique, representing the first such description.

Thiamine Deficiency in Serious Illness

This chapter is entitled thiamine deficiency in serious illness and will examine relatively recent descriptions of the development of thiamine deficiency in various cancers, gastric bypass surgery, etc. Predating these discoveries have been thorough descriptions in prisoners of the development of thiamine deficiency. These cases are different from those in various cancers in that they are clearly based on dietary inadequacies. They are similar however in that these men were seriously ill and have no other counterpart in clinical medicine. Therefore a comprehensive study and review of this prisoner condition is synopsized here (Denny-Brown, D., 1947).

Beriberi

Dr. Denny-Brown served as a consultant in neurology in Southeast Asia immediately following WW2 and was in a position to examine thousands of released prisoners of war. There were various "releases" of prisoners and in one such release there were 650 prisoners, and only 235 of them were sick. Of these there were a surprisingly low number of overt vitamin deficiency cases. There were several cases of beriberi, as well as cases of ataxia, edema, and polyneuropathy. The vast majority had malaria, dysentery, and helminth infestations. The number of cases dying before release from any of the above conditions, including beriberi, is unclear.

In another prisoner release, beriberi was recognized in a number of patients. Peripheral neuritis, edema, and ataxia were all noted; however, the number of these cases was lower than was expected. Other nutritional deficiencies were diagnosed including pellagra. The findings of other neurological deficits such as retrobulbar neuritis, spastic paraplegia, and acrodynia (burning feet) could have been independent of thiamine deficiency. These conditions have been described as a component of beriberi, but the author speculated that they occurred independently.

Most released patients were suffering from a variety of illnesses, some severe, some not severe. In addition, thousands of prisoners died before release, and these may represent many severe cases of nutritional deficiency coupled with other serious illnesses. Nevertheless, these released prisoners had nutritional deficiencies (thiamine) coupled with diseases such as malaria. The interrelationship between these two disorders remains unclear.

D.W. McCandless, *Thiamine Deficiency and Associated Clinical Disorders*, Contemporary Clinical Neuroscience, DOI 10.1007/978-1-60761-311-4_10,

Although thiamine deficiency (beriberi) has been medically described since 1642, thiamine deficiency in serious illness (TDSI) is a relatively new clinical finding. In addition to thiamine deficiency in prisoners, it has been described in patients with breast cancer, hematologic cancers, gastrectomy patients, HIV/AIDS, and others. In some instances the thiamine deficiency is mild and measured largely through laboratory tests. These cases might have no overt symptoms, but the possible contribution of the deficiency to morbidity and mortality is unknown. In other cases, the thiamine deficiency can be overt and severe on its own and produce deleterious effects such as cardiac failure.

Cancer

Another type of thiamine deficiency associated with serious illnesses are cancers treated with 5-fluorouracil. This anticancer agent appears to be an antagonist to thiamine derivatives, and thereby produces thiamine deficiency in some patients being treated with the drug.

One of the first such studies (Basu, T. et al., 1974) of TDSI was of transketolase activity in cancer patients undergoing 5-fluorouracil treatment. Transketolase, a thiamine-dependent enzyme was chosen as an indicator of thiamine deficiency. Other easily obtained test data such as blood pyruvate levels and urinary excretion of thiamine are non-specific. The measurements of blood transketolase activity and the thiamine pyrophosphate (TPP) effect (the increase in transketolase activity when TPP is added to the assay) are, when taken together, excellent measures of relative thiamine deficiency. These two assays were used in this study to measure thiamine deficiency.

Results showed that in cancer patients (breast, respiratory system, bladder, cervix, bowel, etc.) transketolase activity was not statistically significantly changed as compared to controls. When adding TPP to the reaction, however, the values of activity for transketolase were significantly elevated as compared to controls. The elevation was about 65%. To further examine the possible cause of the effect on transketolase, 5-fluorouracil was added to red blood cell hemolysates. Results showed a graded effect in that the transketolase activity was decreased in proportion to added 5-fluorouracil.

Even though the cancer patients were receiving multivitamins, many had significant values for the TPP effect, indicating a significant level of thiamine deficiency. Clinical evidence of thiamine deficiency was absent. However, these data suggest that cancer patients are susceptible to thiamine deficiency. The overall effect of minimal thiamine deficiency in cancer patients is unclear, but bears further investigation.

The effect of 5-fluorouricil is consistent with the concept that it is an antagonist to TPP. The exact mechanism by which 5-fluorouricil acts on thiamine is not clear. Further studies examining how 5-fluorouricil works on the transketolase reaction and on the effects of other antitumor agents are warranted.

In another paper by Basu (Basu, T. and Dickerson, J., 1976), the thiamine status is examined in patients with breast and bronchial cancers. A total of 42 patients with early breast or bronchial cancer who had not yet received chemotherapy or surgery were studied. In these patients, the TPP effect was measured in cancer patients and non-cancer controls. Results showed that the TPP effect was increased over twofold in bronchial cancer and threefold in breast cancer patients. When examined for thiamine, results showed an increase of the vitamin in urine of the bronchial and breast cancer patients. There was no evidence for any nutritional deficiencies in any of the 42 patients examined, nor evidence of malnourishment due to loss of appetite.

The increase in urinary thiamine could be due to a defect in absorption of thiamine in the gut, or it could also be due to a failure to convert thiamine to TPP. The authors state that available data suggest that in these two cancers, there is a decrease in the conversion of thiamine to the coenzyme form, TPP. Further study should determine the exact cause of thiamine deficiency in these illnesses.

Intensive Care Units

In a paper (Cruickshank, A. et al., 1988), a retrospective study was done to examine possible thiamine deficiency in critically sick patients. In this study, 152 patients were divided into two groups: those who died in the ICU within 1 week and those who did not die. There were an equal number of patients in each group. Transketolase activity and the TPP effect were measured in each case.

Results showed considerable variation in transketolase activities within the two groups. The activation of transketolase in RBCs in patients who died was significantly increased. This value is derived from the initial transketolase value, and that measured after TPP is added to the reaction. An increased TPP effect is taken as evidence of thiamine deficiency. A few of the critically ill patients in the ICU continued to have poor transketolase activation values, and died despite thiamine supplementation of 1.2 mg/day.

This study shows a correlation between mild thiamine deficiency and failure to survive in the ICU. The exact cause for this is unclear. Severity of illness seemed to be about equal between groups. The overall nutritional status of these patients could not be fully assessed, but those in the group who failed to survive may have had a lowered status. Of the group who were thiamine deficient and died, seven had hemodynamic deterioration before death and one had neurological complications. None of these showed any pathological evidence of either beriberi or Wernicke's disease. Beriberi can present clinically as either high- or low-output diseases, making diagnosis in otherwise ill patients difficult. The presence of thiamine deficiency in otherwise sick patients may influence the patients' outcome, but exactly how and to what degree is uncertain.

The approximate daily requirement for thiamine is about 1 mg/day, but this is increased in conditions in which metabolic rate might be increased such as in some cancers and other serious illnesses. In parenterally fed patients, the suggested requirement could be as much as 3 mg/day (American Medical Association, 1975). These levels of thiamine supplementation could probably be increased somewhat.

One way to assess this is by performing frequent transketolase assays, but this is not practical. Evidence for clinical improvement using thiamine supplementation is inconclusive, but such additional thiamine is certainly not detrimental, thereby justifying its use in critically ill patients. Further investigation is needed to evaluate thiamine deficiency in otherwise very sick patients.

Hematologic Tumors

In another study (van Zaanen, H. and van der Lelie, B., 1992), thiamine deficiency was diagnosed in patients with fast growing hematologic tumors. Six patients were examined for thiamine deficiency, and among these, two were found. Both patients had severe cardiovascular symptoms. These symptoms were treated with thiamine supplementation, and the symptoms quickly reversed. In neither of the patients was there any evidence of neurological symptoms commonly seen in Wernicke's disease. Neither of these two patients were abusing alcohol.

The authors speculate that because leukocytes have levels of thiamine-dependent enzymes (pyruvate dehydrogenase, alpha-ketoglutarate dehydrogenase, and transketolase), the hematologic tumors have even higher consumption of thiamine, thereby leading to thiamine deficiency. The authors state that this finding in cases of hematologic rapidly growing tumors suggests that clinicians be vigilant for other such cases. The presence of thiamine deficiency in these types of cases should prompt treatment with increased thiamine supplementation. The possible beneficial effects from early recognition of the deficiency and early treatment are stressed.

Gastrectomy

Yet another class of patients have been described with thiamine deficiency (Koike, H. et al., 2001). In this study, 17 patients were examined who developed polyneuropathy following gastrectomy. Patients included in this study had had a total or partial gastrectomy due to ulcers or tumors, and were thiamine deficient and developed a polyneuropathy. The polyneuropathy was confirmed by sural nerve biopsy. Thiamine levels were measured in blood, and transketolase activity was measured in RBCs. Patients with other neurological disorders were excluded from the study, including alcoholics. Sensory and motor conduction studies were performed, and a psychiatric evaluation was done.

Results showed that the first symptoms were weakness in arms and legs in 11 patients. All 17 eventually developed extremity polyneuropathy with the lower extremities being more affected. With time the trunk became involved in five patients. Inability to walk was common to almost all patients. Burning and numbness was present in all 17 patients. Cranial nerves were affected, as evidenced by facial palsies and gaze palsies.

The Wernicke-Korsakoff syndrome could be diagnosed in four patients, and heart failure was diagnosed in ten patients. Edema was present in the legs of most patients. All patients had blood thiamine levels below that of controls. Studies of nerve conduction showed significantly decreased amplitude of compound muscle action potentials as well as sensory nerve action potentials. The density of myelinated fibers in the sural nerves was significantly reduced. All myelinated fibers were affected. Unmyelinated fibers were decreased as well.

Treatment of these patients with 100 mg of thiamine daily had a marked effect. Heart symptoms such as cardiomegaly, arrhythmias, and edema were improved in a few days. Improvement in muscle weakness could be noted in a few days, and the patients' ability to walk returned after 3–6 months of thiamine administration. Overt Wernicke's disease improved with thiamine treatment (not unexpectedly), whereas symptoms of Korasakoff's psychosis remained in one patient for at least 2 years.

Both thiamine levels and transketolase activity levels were decreased in all patients in this study. Thiamine levels proved to be more reliable than transketolase activity levels. Blood levels of most other vitamins were normal or only slightly decreased, but no symptoms of other vitamin deficiencies were noted. The clinical symptoms of these patients were comparable to those seen in beriberi. These included polyneuropathy, cardiac manifestations, and edema. The rapid improvement in symptoms following thiamine treatment further emphasizes the thiamine deficiency characteristics of this disorder.

Gastrectomy was a common procedure in all patients in this study. A few reported cases of thiamine deficiency in gastric resection patients had been reported earlier (Abarbanel, J. et al., 1987). None of the present cases were operated on for obesity, only for ulcers or tumors. Vomiting, common in surgery for obesity, was not a feature of these cases. Previous reports of Wernicke's disease have appeared in cases of gastrectomy (Shimomura, I. et al., 1998). Since thiamine is largely absorbed in the intestine, which is not included in a gastrectomy, the exact mechanism of how thiamine deficiency develops is unclear.

The authors further state that thiamine deficiency symptoms did not appear immediately after surgery, but some did have an acute onset. Because of a possible decrease in eating prior to surgery due to ulcers or tumors, a mild subclinical thiamine deficiency occurred. After surgery the patients might have an increased demand for thiamine as is seen in cases where thiamine deficiency can be precipitated by fever, increased muscular tone and activity, pregnancy, or other causes. Also, it is possible that treatment of tumors with agents such as 5-fluorouracil could also produce thiamine deficiency by inhibiting thiamine activity as described earlier in this chapter.

In another study (Seligmann, H. et al., 2001), the thiamine status was assessed in 14 patients with B-chronic lymphocytic leukemia (CLL). This leukemia was diagnosed based on the National Cancer Institute guidelines. Excluded from this study were alcoholics, malnourished patients, malabsorption syndromes, patients with vomiting, prolonged fever, liver disease, etc. In the 14 patients with CLL, and 14 controls, thiamine status was assessed by erythrocyte transketolase activity and by the TPP effect.

Results showed thiamine deficiency, as measured by transketolase and the TPP effect, in 5 of the 14 CLL patients. None of these five showed any overt clinical signs or symptoms. Since the study had a rigorous exclusion criteria, it would seem that CLL played a role in the thiamine deficiency. This could be due to an increased thiamine requirement by tumor cells, increased whole body requirements, or decreased TPP. In keeping with these possible mechanisms was the finding that the thiamine deficiency was present in patients with advanced disease. While the five patients with thiamine deficiency were asymptomatic, the presence of a "mild" thiamine deficiency in the CLL disease process is yet to be determined.

It is noted by the authors that treatment of this subclinical thiamine deficiency in cancer patients should be approached carefully. For example, ribose synthesis coming from the pentose phosphate pathway (a thiamine-dependent pathway) could be increased by thiamine treatment. Increase in ribose for nucleic acid synthesis could possibly be a promoter of tumor cell growth in various cancers. Thus, excess thiamine therapy might promote tumor growth, and overall aid the disease process.

Previous studies had shown that cancer treatment was associated with thiamine deficiency, and the exact cause was speculative. Additional studies examined the effect of various antitumor drugs on thiamine metabolism. In one such study, the neurological signs and symptoms of 5-fluorouracil were described (Moore, D. et al., 1990b).

5-Fluorouracil

5-Fluorourcil is an antimetabolite used widely as an antitumor agent against a number of solid tumors. It is known that 5-fluorouracil can produce neurological symptoms, although this is not common. The paper presents two cases of 5-fluorouracil neurotoxicity. Both patients had genital tract malignancies, which were treated with 5-fluorouracil.

Within a few days of the onset of 5-fluorouracil treatment, signs and symptoms such as weakness, slurred speech dysarthria, nystagmous, and ataxia were noted. In one case, computed tomography was normal. Following the diagnosis of 5-fluorouracil, the drug was discontinued and in one case, symptoms were greatly improved in a few days. In the other case following diagnosis, thiamine therapy was instituted, and symptoms were completely reversed in 3 days.

A literature review by these authors found the neurological symptoms of ataxia, slurred speech, and cerebellar ataxia in patients receiving 5-fluorouracil to be only 2–5%. The symptoms of 5-fluorouracil neurotoxicity are fully reversible with cessation of 5-fluorouracil therapy and/or treatment with thiamine. This 5-fluorouracil disorder has an acute onset, and is present in the absence of other possible contributors such as brain lesions, diabetes, or fever.

The mechanism by which 5-fluorouracil produces neurological symptoms is unclear. One metabolite produced when 5-fluorouracil breaks down is fluorocitrate, which is in turn a TCA cycle inhibitor. It has also been shown that 5-fluorouracil can block the conversion of thiamine to TPP, the active form of thiamine. Since nutritional deficiencies may be borderline in cancer patients, the added burden of 5-fluorouracil may serve to push vulnerable patients into frank thiamine deficiency. Further 5-fluorouracil toxicity studies should shed light on the mechanisms of neurological deterioration associated with its use.

In another study, the diuretic drug furosemide used to treat congestive heart failure was examined for its ability to produce thiamine deficiency (Seligmann, H. et al., 1991). Patients were divided into two groups: group A with 23 patients with chronic heart failure receiving furosemide and group B with 16 control patients without congestive heart failure and not receiving any diuretics. All patients received the same diet, and thiamine status was evaluated by measuring urinary thiamine output, and measuring the blood transketolase activity, and the TPP effect.

Results showed an increased TPP effect in 21 of the group A patients and only 2 of the group B patients. Urinary excretion of thiamine was greater in the group A patients as compared to the control group. Transketolase activity measurements were similar between group A and group B.

Thiamine deficiency can cause cardiovascular and neurological signs and symptoms. These signs and symptoms are rapidly reversed by thiamine administration. Mild thiamine deficiency can exist unnoticed for a long time and be measured by such tests as increased thiamine in the urine, high blood lactate and pyruvate levels, and decreased transketolase activity and increased TPP effect. It has been found that the diuretic furosemide produces an increase in urine thiamine and an increased TPP effect, both indicators of thiamine deficiency.

The mechanism of furosemide toxicity is not clear. The diuretic does cause anorexia, hyponatremia, and hypomagnesemia. These three alterations in metabolism could change normal thiamine metabolism. The initiation of thiamine therapy resulted in the rapid return to normal of the increased TPP effect. Thiamine treatment also served to improve cardiac contractility. Thus, thiamine deficiency should be diagnosed and treated since a failing heart can be adversely affected by thiamine deficiency, as seen in beriberi heart disease.

The drug omeprazole (prilosec) has widespread use as a drug for treating acid-induced inflammation and ulcers in the stomach and duodenum. Data show that omeprazole is detrimental to the normal function of thiamine metabolism (Nixon, P. et al., 1992). In this study, the effect of omeprazole was evaluated in vitro on the activity of the two thiamine-dependent enzymes transketolase and pyruvate decarboxylase.

Transketolase was prepared from human red blood cells. Possible omeprazole inhibition was evaluated by adding the drug to media containing apotransketolase and measuring activity spectrophotometrically. Pyruvate decarboxylase was prepared from the bacteria *Z. mobilis* and as above, measuring activity spectrophotometrically.

Results showed that both thiamine-dependent enzymes transketolase and pyruvate decarboxylase were inhibited by omeprazole. The results indicate that omeprazole is an analog of thiamine, and inhibits two thiamine (TPP) requiring enzymes, transketolase and pyruvate decarboxylase. This drug therefore may have a detrimental affect on patients who may at ready have threatened thiamine reserve due to anorexia.

While thiamine deficiency can occur in cancer patients, and can be deleterious, in another view, the over administration of therapeutic thiamine may be conducive to tumor growth (Boros, L. et al., 1998). It is well known that the pentose phosphate pathway is a key source of ribose production, and therefore RNA and DNA synthesis depend on a well-functioning pentose phosphate pathway. The rate-limiting step in this pathway is the enzyme transketolase, a TPP-requiring enzyme.

There are clearly several mechanisms whereby cancer patients may develop thiamine deficiency (see above). This thiamine deficiency may be severe in that Wernicke's, beriberi, lactic acidosis, etc., may develop, further complicating the cancer treatment. It has been shown that ribose production in pancreatic adenocarcinoma cells proceeds through the pentose phosphate pathway and that the administration of the antithiamine metabolite oxythiamine inhibits tumor cell proliferation significantly (Boros, L. et al., 1997). The tumor cell inhibition is demonstrable both in vitro and in vivo.

Since the treatment of even mild thiamine deficiency in cancer patients is usually achieved by an administration of several times the minimal daily requirement for thiamine, this treatment regime might be conducive to tumor growth. This excess treatment of cancer patients with thiamine could increase ribose production, thereby rendering treatment of cancer patients with anticancer drugs less affective.

The finding of thiamine deficiency in patients with gastrectomy has been described (Sekiyama, S. et al., 2005). This paper describes a patient who had a total gastrectomy for a gastric signet ring cell carcinoma. The patient felt that the cancer might have been related to diet, so stopped eating meat, chicken, fish and milk, and ate only rice, noodles, cabbage, onions, radish, etc. A couple of years later, the patient had lower extremity edema, plus tingling and pain in the feet.

A neurological exam showed the patient had edema, lower extremity weakness, and absence of deep tendon reflexes. Lower extremity hyperesthesia and paresthesias were present. A diagnosis of beriberi was confirmed based on serum concentrations of vitamin B1 and B12, which were decreased from normal. Thiamine therapy was initiated (75 mg/day), and a proper diet was started. The patient's signs and symptoms improved, and she made a full recovery.

This patient had a gastrectomy which acted to decrease the length of duodenum available for thiamine absorption. In addition, the diet she adopted following surgery was inadequate in many ways, including thiamine content. Other cases of thiamine deficiency have been associated with gastric resection, and this case emphasizes other problems which can coexist, such as poor diet and alcoholism. The poor diet is frequently seen in cases of gastrectomy because patients erroneously think that a protein-rich diet may have been responsible for their disease. This patient also had a vitamin B12 deficiency, which may also have contributed to her clinical picture.

Emphasizing the risks for certain types of gastric bypass surgery such as the "Roux-en-Y" type, a recent study examines neurological complications (Juhasz-Pocsine, K. et al., 2007). In this study, nearly one half of patients showed ataxia/walking difficulty and loss of sensation in the lower extremities. Some patients may have had brain involvement. In other patients, the neurological complications appeared to be limited to the peripheral nerves. These neurological signs often develop from 1 to 2 months after surgery and when diagnosed can be effectively treated with thiamine in a proper dose.

Wernicke's Disease and Gastrectomy

Wernicke's encephalopathy can develop many years after gastrectomy (Karapanayiotides, T. et al., 2006). This brief article describes a patient who developed unsteady gait, vomiting, and dizziness. Three years earlier he had experienced polyneuropathy. He had been drinking 1–2 drinks per day, but was not an alcoholic. Twenty-eight years earlier, he had a partial gastrectomy due to a peptic ulcer. He developed nystagmus, abducens palsy, confusion, stupor, and continued to vomit following surgery.

MRI showed lesions in the floor of the fourth ventricle, periaqueductal gray, and hypo and medial thalami. These images were suggestive of Wernicke's disease, and thiamine treatment was initiated. Three hours after thiamine therapy was begun, symptoms (abducens palsy) began to abate. One month later, there was minimal nystagmus and amnesia. The authors state that the poor diet, alcohol consumption, and vomiting precipitated this case of Wernicke's disease many years after the gastrectomy. Thiamine therapy resolved most signs and symptoms.

In a brief case presentation (Boniol, S. et al., 2007), a patient was described with malaise, large weight loss (50 lbs), and diffuse abdominal pain. He was jaundiced, had lymphadenopathy, and vomiting. He was ultimately diagnosed with non-Hodgkin's lymphoma. Nineteen days after initial chemotherapy, he complained of difficulty in walking and blurred vision. Exam showed nystagmus, but MRI was unremarkable. Thiamine therapy was initiated, and within 48 hours all neurologic symptoms were gone. Ten weeks later he was still free from neurological symptoms. This represents a case of Wernicke's disease in a non-alcoholic patient who developed thiamine deficiency through decreased oral intake and increased cell turnover. Since the diagnosis of Wernicke's disease can be largely clinically based, this possibility, although rare, must be considered since prompt thiamine therapy can quickly reverse symptoms.

In the mid to late 1980s, several clinical papers were published describing thiamine deficiency like lesions (Wernicke's disease) shown by autopsy findings in patients dying of AIDS. The first descriptions seem to have been published by Foresti, V. et al. (1986), and by Rosenberg (Rosenberg, S. et al., 1986). A diagnosis of Wernicke's disease was only occasionally suspected before death; most cases were diagnosed by autopsy findings. Lesions were found in the usual sites for Wernicke's disease: the pons, thalamic nuclei, and mammillary bodies. Occasional lesions were noted in other brain areas. The lesions were hemorrhagic in nature, and microscopically, there was neuronyl cell loss and demyelination.

HIV/AIDS

In an attempt to determine if AIDS might be associated with thiamine deficiency, 39 patients were studied for the presence of thiamine deficiency (Butterworth, R. et al., 1991). In this study, the AIDS patients were studied as regards their "TTP effect" status. This test involves measuring the thiamine-requiring enzyme transketolase both with and without added TPP. An increase of 15% or more of transketolase activity after TPP addition is taken as indication of thiamine deficiency.

In an initial screening of the 39 AIDS patients, six were found to be thiamine deficient, and in another screening 6–9 months later found three additional patients with a TPP effect greater than 15%. The total number of thiamine deficient patients was nine, nearly 25%. In none of these positive patients was there a history of alcohol abuse, and none had any overt Wernicke's disease symptoms.

The association of thiamine deficiency and Wernicke-like lesions in AIDS patients is most likely due to the extreme wasting and poor nutritional status of AIDS patients. Indeed, thiamine deficiency had been documented in several cancers and in gastric bypass surgeries (see above, this chapter). In cases of thiamine deficiency in cancers, as in AIDS patients, the degree of deficiency may be mild and subclinical. However, the actual effect of thiamine deficiency on the course of AIDS remains to be assessed.

Since diagnosis of thiamine deficiency in AIDS patients during life is difficult, it behooves the health care worker to be cognizant of the possibility that thiamine deficiency could be present (25% of cases), and if suspected, initiate treatment. The paper of Butterworth et al. is important in that it shows beyond doubt that thiamine deficiency is present in a significant number of AIDS patients, and actually calls for thiamine treatment in all newly diagnosed AIDS cases. These data have been replicated by other investigators, who found a very similar incidence for thiamine deficiency as did Butterworth (Muri et al., 1999).

Other problems associated with thiamine deficiency in AIDS patients may occur. One such problem is a severe lactic acidosis which has a low frequency of occurrence in AIDS patients treated with nucleoside analogue reverse-transcriptase inhibitors (NARTIs). The use of NARTIs in AIDS patients leads to a mild lactic acidemia in about 20% of patients. However, in a much smaller number of AIDS patients (1%), there is a complication of an abrupt severe lactic acidemia. Symptoms from this may include malaise, vomiting, dyspnea, asthenia, etc. Lactate levels may elevate to over 6 mmoles/L. The mortality rate for this development may approach 50% (Calza L. et al., 2005; Walker, V., 2004).

MRI studies of AIDS patients who develop Wernicke-like symptoms show abnormal symmetric hyperintensities in the mammillary bodies, periaqueductal gray matter, and the inferior colliculus, so characteristic of Wernicke's disease (Tattevin, P. et al., 2006).

Treatment is supportive, and an immediate cessation of NARTI treatment. Many cases have in common the inclusion of stavudine in the treatment regimen. The mode of action may involve the fact that these compounds are inhibitors of gamma polymerase, an enzyme involved in the replication of mitochondrial DNA. Other potential mitochondrial toxins such as valproate and acetylsalicylic acid must also be withdrawn. Vitamin therapy can be started, and one case was successfully treated with thiamine (Schramm, C. et al., 1999). This treatment of NARTI-induced severe lactic acidosis with thiamine reinforces the concept that vitamin deficiency is a vital feature of this relatively rare complication of NARTI therapy in AIDS patients. The treatment of this patient with thiamine reversed the hyperlactic acidosis within hours. Possibly, high doses of vitamins should be given to all patients showing this life-threatening development of hyperlactic acidemia.

Malaria

Another area of serious illnesses in which thiamine deficiency has been shown to have a possible detrimental effect is malaria (Krishna, S. et al., 1999; Mayxay, M. et al., 2007). Malaria is, of course, a serious health problem in many parts of the world such as sub-Saharan Africa, parts of South and Central America, and in the Western Pacific, where there are as many as 2 million cases and 10,000 deaths per year.

Perhaps the first to make a connection between malaria and thiamine deficiency was Krishna et al. (1999). In the Lancet they described studies in which the thiamine status was assessed in patients with malaria, and in controls. Results showed that 50% of patients with severe malaria and 20% with mild uncomplicated malaria were also thiamine deficient as compared to control non-malaria patients.

Whether thiamine deficiency contributes to severe malaria or whether malaria exacerbates thiamine deficiency is unclear. It is certain that the combination may speed up the malaria condition, contributing to the morbidity and mortality. The authors suggest that thiamine treatment should be combined with anti-malaria therapy in order to increase treatment outcomes in a positive manor.

A recent paper (Mayxay, M. et al., 2007) examined the occurrence of malaria and thiamine deficiency in the same patient. Severe falciparam malaria can be associated with metabolic acidosis (Dondorp, A. et al., 2004). Malaria is one of the several conditions which may increase glucose flux, thereby raising the need for thiamine in its coenzyme form. Based on these observations, 310 patients were studied as to their thiamine status. A follow-up assessment of the patients' status was performed 42 days later, and the results compared. Thiamine status was evaluated by measuring the activity of the thiamine-requiring enzyme transketolase.

Results showed that 12% of the patients initially tested showed a severe level of thiamine deficiency based on the transketolase assay. Forty-two days later, only 3% of patients were thiamine deficient. On hospital admission, patients were treated with anti-malarial drugs (chloroquine, sulphadoxine-pyrimethamine, and methloquine). Patients were also supplemented with a multivitamin tablet daily containing 1 mg of thiamine. The decrease from 12% at admission to 3% forty-two days later in thiamine status suggests that the thiamine treatment did in fact decrease what was a true thiamine deficiency.

Since younger red blood cells may have higher transketolase levels, and malaria tends to destroy young red blood cells, the estimates of transketolase activity may be underestimated. This study was done in Laos, where polished rice is consumed. Also betel nuts are chewed, and betel nuts contain a thiaminase. So it is highly possible that these people might have compromised thiamine metabolism. Nevertheless none of the patients had any overt clinical evidence of beriberi, or any Wernicke-like symptoms.

The fact that malaria elevates body temperature may also act to increase thiamine deficiency. The authors call for further studies to examine the incidence and morbidity/mortality of thiamine deficiency superimposed on malaria and other acute infections. Otherwise, it seems simple enough to institute thiamine therapy in all cases of malaria. Further education is essential regarding consumption of polished rice, chewing of betel nuts, etc., in order to irradicate once and for all these outbreaks of thiamine deficiency by themselves, and in association with other serious illnesses.

Amyotrophic Lateral Sclerosis

Because of some similarities between inherited ataxias and amyotrophic lateral sclerosos (ALS), thiamine and thiamine monophosphate levels were measured in plasma and CSF in ALS patients, alcoholics, and controls. The ALS patients' diagnosis was confirmed clinically, by electromyography, and by muscle biopsy (Poloni, M. et al., 1982).

Results showed that thiamine and thiamine monophosphate were decreased in plasma to the same extent, rendering the ratio of thiamine/thiamine monophosphate unchanged. In CSF, the decreases were not equal, such that the thiamine/thiamine monophosphate ratio was nearly doubled.

The authors speculate that this finding in CSF is unique in that other neurological conditions show an equal decrease in thiamine and in thiamine monophosphate yielding an unchanged ratio. A selective decrease in thiamine monophosphate raises the ratio. The possibility of a decrease in numbers of neurons in ALS patients is discussed but dismissed. The authors suggest that the diminished thiamine monophosphate could be due to a decrease in thiamine pyrophosphotase activity, leading to reduced enzymatic (protein) synthesis in nerve cells seen in ALS.

A separate paper by the same group (Poloni, M. et al., 1986) reconfirms the previous study in a larger number of patients. In this study, 50 patients with ALS and 37 controls had CSF drawn, and thiamine and thiamine monophosphate and the ratio were measured.

The findings were as in the previous paper. A selective decrease in thiamine monophosphate resulted in an elevated thiamine/thiamine monophosphate ratio. This finding is almost unique to ALS. The mechanism of the selective decrease is not clear. The decrease is not due to thiamine deficiency per se, nor is it seen in most other neurological disorders such as dementia or Wernicke's disease. The authors speculate that there is a defect in the production of thiamine monophosphate in patients with ALS.

Alzheimer's Disease

There is some evidence to suggest changes in thiamine and thiamine-dependent enzymes in Alzheimer's disease (Gibson, G. et al., 1988; Rao, V. et al., 1993). In the second of the above studies, thiamine diphosphatase was measured in frontal and temporal cortex in eight confirmed Alzheimer's patients, and controls. Results showed a 62% reduction of enzyme in the temporal cortex, and a 28% decrease in the frontal cortex, as compared to controls.

These results are in agreement with similar results from other laboratories showing decreased thiamine-dependent enzyme activities in erythrocytes and fibroblasts in Alzheimer's patients. Collectively, these results suggest that altered thiamine metabolism is a key feature of Alzheimer's disease and that altered metabolism is not limited to the CSF. How the findings of changes in thiamine diphosphatase and alterations of TPP-dependent enzymes are related is not clear. Electron microscopic studies have shown that Alzheimer's patients have decreased synaptic densities, which could relate to the observed neurochemical results.

The above studies have been confirmed (Gold, M. et al., 1998) in a study which looked at plasma and erythrocyte thiamine levels in a group of Alzheimer's disease, and Parkinson's disease patients. Results showed that the Alzheimer's disease patients had lower plasma thiamine concentration than did the Parkinson's patients. These data confirm previous findings that thiamine deficiency occurs in the plasma of Alzheimer's patients. The Parkinson's group served as a control group.

A study directly measuring the thiamine-dependent enzymes pyruvate dehydrogenase, alpha ketoglutarate dehydrogenase, and transketolase activities in temporal cortex showed effects in Alzheimer's disease. The results were obtained from six Alzheimers patients and eight controls. Results showed a 70% decrease in pyruvate dehydrogenase activity and a 70% decrease in alpha-ketoglutarate dehydrogenase activity. Transketolase was decreased in temporal cortex by 52%. These results are in keeping with the concept that alteration in key enzymes in energy metabolism may be playing an important role in the pathogenesis of Alzheimer's disease. Further studies might examine energy metabolite levels in this disorder.

An interesting animal model of Alzheimer's disease has been developed in which thiamine deficiency is produced, and the pathology of plaque formation is studied (Karuppagounder, S. et al., 2008). A mouse model of Alzheimer's disease developed in Tg 19959 transgenic mice was utilized. These mice over express a double mutant form of the amyloid precursor protein. Results showed that thiamine deficiency augmented amyloid plaque development. The areas affected in the cortex, hippocampus, and thalamus increased by up to two times. Thiamine deficiency increased AB1-42 levels, B-CTF levels, and B-secretase levels significantly.

These results demonstrate, in a mouse model, that thiamine deficiency increases plaque development and increases key protein levels. This finding is important in that it is direct evidence of a deleterious effect of thiamine deficiency. Some form of treatment of this defect of thiamine metabolism in Alzheimer's disease should be explored.

World Health Concerns

In the United States thiamine deficiency is relatively rare, and when present is usually associated with chronic alcoholism and Wernicke's disease. In the world, however, the picture of thiamine deficiency is much different. For example, the reader is directed to a chapter in this monograph on an outbreak of thiamine deficiency in Nigeria called African Seasonal Ataxia. In these cases, thiamine deficiency was caused by eating beetle larvae which contained a thiaminase. Many hundreds of cases were reported. Other outbreaks of thiamine deficiency have occurred from eating white cereals or from eating polished rice. Outbreaks of thiamine deficiency have recently been associated with food ration distributions. These food aid distributions usually contain food stuffs not well balanced. When consumed by people already semi-deficient, overt thiamine deficiency can appear quickly.

In fact, a "subclinical" thiamine deficiency afflicts large numbers of the world's population who subsist on marginal or submarginal thiamine intake. These people are then prime targets for developing overt thiamine deficiency given the right circumstances, and major epidemics may develop. It is sometimes difficult to predict these outbreaks due to the lack of a simple reliable test for subclinical thiamine deficiency.

There is a low capacity for the human body to store thiamine, the liver being the main site. Overt thiamine deficiency can develop in as little as 2–3 weeks in subclinical patients when an incomplete thiamine diet is administered. High-carbohydrate diets and low-thiamine intake precipitate thiamine deficiency and can lead to the easily detectable thiamine deficiency. Anorexia, a common symptom in thiamine deficiency, can be considered beneficial in that it reduces the deleterious effect of high carbohydrate meals. Another at-risk group is people who are at the margin of thiamine deficiency and suddenly have increased needs. Examples include pregnancy and lactation, people who have heavy exertion, and those with other illnesses such as liver disease, cancer patients, and in situations in which absorption is compromised such as in GI disease.

The symptoms of mild thiamine deficiency are hard to recognize and are often attributed to other disorders. These early symptoms are quickly and easily reversed by thiamine treatment given proper nutrients and the knowledge to administer them. Deficient biochemical thiamine states have been associated with a variety of other

D.W. McCandless, *Thiamine Deficiency and Associated Clinical Disorders*,
Contemporary Clinical Neuroscience, DOI 10.1007/978-1-60761-311-4_11,

disorders, including reduced growth in children, lactic acidosis, renal dysfunction, arrythmias, SIDS, endocarditis, cancers, and HIV/AIDS.

There were large outbreaks of thiamine deficiency in Southeast Asia around the early 1900 s which were related to large-scale production of milled rice and large-scale distribution. The milled rice was cheap and popular. Milled rice is a significant risk factor, which is increased in pregnancy, fever, and muscular activity. Other risk factors reported (Rolfe et al., 1993) in Gambia included alcohol consumption, fever, pregnancy, diabetes, and dysentery. Milled rice constituted nearly one half of all caloric intake. Intense activity during the agriculture/rainy season was also a risk factor for developing overt thiamine deficiency. In some cultures, eating thiaminase containing larvae can quickly lead to thiamine deficiency.

Breastfeeding women with signs of thiamine deficiency usually were not eating as well as those without symptoms. These women were eating well under the minimum daily requirement of thiamine. In addition many women chewed betel nuts, which contain a thiaminase. Thiamine-deficient mothers have thiamine levels in their breast milk insufficient to provide enough thiamine to fulfill the minimum daily requirement for infants (Valyasevi, A. and Vimokesant, S., 1968). Symptoms in infants include dyspnea, GI dysfunction, vomiting, crying, cardiac signs, etc. In infants, these signs usually progress quickly to stupor and death.

The minimum daily requirement for thiamine in adults is about 1.2 mg per day. This number varies and depends in part on the percent of intake of carbohydrates plus risk factors. Other factors also influence thiamine requirements such as ambient temperatures, increasing age, body weight, and overall health.

Thiamine is widely distributed in food stuffs, both plant and animal, but most contain low quantities. In addition, preparation (polishing) of rice, consumed in large quantities world-wide, removes nearly all thiamine. Pasteurization of cow's milk removes some thiamine. The consumption of refined cereals as well as products made from these white cereals (bread) does not contain enough thiamine to prevent illness and death. Other sources such as intestinal bacteria, which synthesize thiamine, and intestinal fermentation add to thiamine available to humans.

Strategies need to be developed and implemented in order to supplement thiamine availability in third world countries. Possible approaches include providing food rations that contain a variety of food stuffs, including fruits and vegetables. Another possible solution is to reduce the milling of rice. The idea of fortification of food stuffs – adding thiamine to the milled rice, for example – has been around for many years, but has not been widely adopted by manufacturers of rice or wheat flour. The fortification process is not easily achieved, and what is available is a compound called vitarice, which is added in the home. Vitamin supplement can also be taken in popular foods such as biscuits or cookies enriched in vitamins. Supplementation using vitamin tablets is logistically difficult and expensive.

Loss of thiamine can be lowered if people can be educated and decrease the milling and polishing of rice prior to consumption. Cooking rice in lower volumes serves to preserve thiamine. Timed cooking of both rice and vegetables serves to lower loss of vitamins. Also, eating a prepared meal as soon as it is ready lowers loss of vitamin. Finally, as mentioned earlier, many substances contain a thiaminase, which can breakdown thiamine and produce symptoms in only a few hours. An earlier chapter in this book details the thiaminase in larvae responsible for African Seasonal Ataxia. Much world wide education is required.

Lest these comments be thought of as trivial or confined to just a few cases, major outbreaks of thiamine deficiency have occurred. There are many examples of large outbreaks. In northern and eastern areas of Thailand where rice is consumed in large quantities, the prevalence of thiamine deficiency may always be from 5 to 8% in adults. This translates to thousands of cases. In Nepal in a 1-year period of 1993–1994, 12,000 cases were suspected. In one refugee camp in Nepal, as many as two cases were reported per 10,000 people per day. Many more cases may go undiagnosed because of cultural reasons for not seeking medical advice. It is possible that subclinical cases around the world, and in the US, are frequently undiagnosed, yet contribute to overall morbidity.

In a recent study (Fattal-Valevski, A. et al., 2005), an outbreak of thiamine deficiency was described in Israel. In this study, an initial case was described in a 5-month-old infant who presented with nystagmus, ophthalmoplegia, and vomiting. Some form of Wernicke's disease was suspected, and thiamine was administered. Improvement was noted within several hours. Eight other infants were eventually admitted with similar symptoms. Findings included elevated blood transketolase (TTP effect) to "severe" levels. Lactate and pyruvate levels were elevated in blood and CSF in some of the patients. Three patients had neurological symptoms, and eight of nine had evidence of GI distress, including vomiting, diarrhea, and in one case, intussusception. An MRI of one patient showed symmetric signals in the basal ganglia, mammillary bodies, and periaqueductal gray matter. This patient had a rapid downhill course and became comatose. Thiamine was administered and improvement noted 24 hours later. Many patients had infectious processes, including pneumonia, *Salmonella* infection, upper respiratory infections, etc. It was finally determined that all cases had consumed a soy-based formula (Remedia Super Soya 1) which had essentially no thiamine content. This formula was quickly removed from the market and warnings distributed. Estimations were that only about 20 cases developed probably because the formula was only marketed a few months before outbreaks developed. This outbreak took place in Israel, and it was suspected that between 1500 and 5000 infants had consumed the formula.

The early clinical symptoms were at first attributed to the infections found in these infants. The neurological symptoms of nystagmus and ophthalmoplegia were indicative of brain stem involvement and suggested a severe condition. There are similarities between these cases of pure thiamine deficiency, which is treatable, and Leigh's disease, which is currently not treatable. Once again it is noted that administration of dextrose and other carbohydrates may cause the rapid depletion of remaining thiamine, neurological symptoms. Early diagnosis and treatment with thiamine can produce precipitating dramatic reversal of symptoms and once again points out the presence of a biochemical lesion. In one patient in this study, thiamine treatment did not start until the third day, and some neurological sequelae remained long after treatment was started.

One problem with diagnosis is that the transketolase TTP effect assay is not available in many hospitals. The lactate/pyruvate assays are not always specific for thiamine deficiency. These factors plus the relative rarity of thiamine deficiency in developed countries may slow down diagnosis. In addition, all patients had infectious processes, which contributed to slower diagnosis.

A retrospective study of an outbreak of beriberi in a prison in the Ivory Coast was recently published (Ahoua, L. et al., 2007). In this study, symptoms such as bilateral leg edema, dyspnea, ataxia, and paresthesias were taken as evidence for possible beriberi. Seven hundred and twelve cases were identified, 115 were labeled "probable", and 597 "definite." The overall prison rate was 14%, and among beriberi patients, the fatality rate was 1%.

Outbreaks of thiamine deficiency (beriberi) in prisons are not uncommon, and as in this case, the main diet was composed of rice. Additional food sources were a prison garden and food brought in by families. These extra activities around the time of the outbreak had been suspended because of political crises both inside and outside the prison.

The outbreak in this prison was recognized early and thiamine treatment initiated. This resulted in 99% of the patients benefiting from the treatment. Only seven patients died, yielding a fatality rate of 1%. The outbreak also resulted in a more equitable distribution of food from the kitchen such that each prisoner received more food and each received the same portions. Malnourished prisoners received two times the normal ration.

This paper reported a large outbreak of 712 cases in a prison population of over 5000. The outbreak was higher in some buildings than others due to unequal food distribution. It was also noted that prisoners with a history of cholera had a higher chance of developing beriberi. The author concludes that it is important to assess the quality and quantity of food rations in prison settings. Mild cases of beriberi may well go undiagnosed by healthcare providers. They should be alert for signs and symptoms in their population of patients.

Another outbreak of beriberi has recently been described in commercial fishermen in Thailand (Doung-ngern, P. et al., 2007). In this outbreak, the crew of a large fishing vessel (28 members) were those who had symptoms such as lower extremity edema, dyspnea, chest pain and discomfort, and extremity numbness. Hospital laboratory tests included transketolase and the TTP effect test.

Results from this outbreak study were that 15 out of 28 developed symptoms of beriberi, and of these, 2 died. All affected patients were hypertensive, nine had pitting edema, five had ascites, etc. Three who had laboratory tests were all positive for thiamine deficiency. Four of the patients had X-ray evidence of cardiomegaly and pulmonary edema. The two patients who died were the first two who developed symptoms, and each died 2 days after onset of symptoms. The next group of patients was treated with thiamine and dramatically improved. Diets of affected sailors consisted mainly of fish and rice.

The authors point out that follow-up to this outbreak was difficult because crew members dispersed when the ship docked. The authors tried through newspapers to remind people that the risk for developing beriberi was still persistent. It is important for healthcare providers to be aware of the signs and symptoms of beriberi and to develop plans for the nutritional well-being of sailors and crew of vessels carrying their own limited food supplies.

Another place where mild thiamine deficiency surely is associated with serious illness would be in Africa among the millions with HIV/AIDS. Even the toll of HIV/AIDS alone is difficult to estimate because of under-reporting. The recognition of the possibility that more than 25% might also be thiamine deficient probably has not been considered. But as treatment develops, the co-presence of other diseases such as malaria and thiamine deficiency must be considered. Remember that millions of indigenous people live at substandard dietary intake.

Epilogue: Future Prospects

This final chapter is a collection of thoughts, questions, prospects, and speculations regarding some aspects of the previous chapters. Some issues raised in this chapter may be trivial or have already been answered. Others will hopefully stimulate discussion and perhaps lead to further investigations into the pathogenesis of thiamine deficiency and associated disorders. These comments are arranged in the same sequence as the chapter headings.

Animal Models of Thiamine Deficiency

One seemingly unanswered question remains the mechanism by which thiamine deficiency produces such striking focal bilateral symmetric lesions. This question is unanswered both in experimental animal models and in human examples of thiamine deficiency such as Wernicke's disease.

In experimental animal models, the model most likely to generate useful data is that of the pure thiamine-deficient diet. The combined use of diet and antimetabolite (pyrithiamine) to produce overt thiamine deficiency, although faster than diet alone, clearly produces results which are somewhat at variance with the diet model alone. These differences include lesions in different areas, and neurochemical results which are at variance with the more chronic model produced by thiamine-deficient diet alone. Certainly Wernicke's disease is a chronic outcome, and any animal model should be as close to the human counterpart as is possible. The diet alone model takes 4–4.5 weeks, a greater percentage of the lifespan of rats than anti-thiamine models.

Several investigators have attempted to discern the regional mechanisms in thiamine-deficient animals responsible for the highly selective lesions. There also have been studies centered on whole brain assays. These studies, by definition, include in the assay material large quantities of tissue which may be non-affected – at least by lesion placement. Inclusion in the sample of non-affected tissue dilutes out the affected tissue, rendering results much less interpretable. If a brain region such as a portion of the pons weighing 30 mg in a rat is homogenized with the rest of the brain weighing 1.2 g, any meaningful change in the pons will certainly be lost in

D.W. McCandless, *Thiamine Deficiency and Associated Clinical Disorders*, Contemporary Clinical Neuroscience, DOI 10.1007/978-1-60761-311-4_12,

data from whole brain. It will be lost through dilution, through animal (biological) variation, and through methodology resolution.

One other frequently omitted feature of a good thiamine deficiency study is the inclusion of controls called "pair-fed controls." These animals are fed the exact amount of food consumed by the thiamine-deficient animals. Each day a weighed amount, for example, 20 g, is given to each individually caged thiamine-deficient animal. Next day at the same time, the leftover food, both in a cup and dropped, is weighed. This determines the amount consumed by the deficient animal. This amount, for example, 3 g, is then offered to the pair-fed control in a one-to-one pairing. This provides a semi-starvation control for each thiamine-deficient animal. This accounts for the effects of starvation. Although a 30-day period for the development of symptoms and pair-feeding is more labor intensive, the data derived from such a study is much more meaningful.

Studies directed at the mechanisms underlying such highly specific localized changes in brain regions such as the thalamic nuclei or cerebellar layers need to be performed. Assays that can be coupled to NAD/NADP have been developed over the last 50 years by Passonneau and Lowry (1993). These methods allow measurement of metabolites directly in single cells. Use of these methods will yield clear answers as to the mechanisms producing highly regional changes in neurochemistry in thiamine deficiency. Remember, studies can be done in experimental animals which cannot be done in man. Why not use the best assay techniques and the animal models most applicable to man?

Remember that animal models represent a unique opportunity to control and manipulate the conditions of the experiment. Carefully controlled experiments are possible, unlike gathering data from human subjects. Animal experiments can be designed very precisely, and controls made nearly perfect. Why compromise this opportunity by not using the best possible model? Or by not having the appropriate controls, or not obtaining results from the most selective brain region? Use direct biochemical methods of analysis designed to answer valid questions, not methods which happen to be available, then try to link results through many steps to possible mechanisms of neuropathology.

Anesthetics prior to sacrifice, or actual method of sacrifice, sometimes mandated by animal-care committees, often produce their own modification of neurochemicals. It is sometimes difficult to determine these possible effects due to the nature of the regulation. Scientists should take a more active role in the sense that arbitrary creation of these animal-care regulations, many of which have acted to significantly increase the cost of medical research, also may compromise data validity.

Another advantage of animal experimentation is to control the killing method such that compounds to be measured are fixed in as close to in vivo as is possible. For example, when labile metabolites are to be measured, rapid inactivation of metabolic processes is essential. An example of how this might lead to spurious data is shown in Fig. 1. Imagine that at 0 seconds levels of a labile metabolite are different than control levels and that their decay rates are different. Note that if freezing does not occur until 30 seconds, then levels appear the same. Some years ago, much time and effort was expended developing rapid killing methods. Devices such as

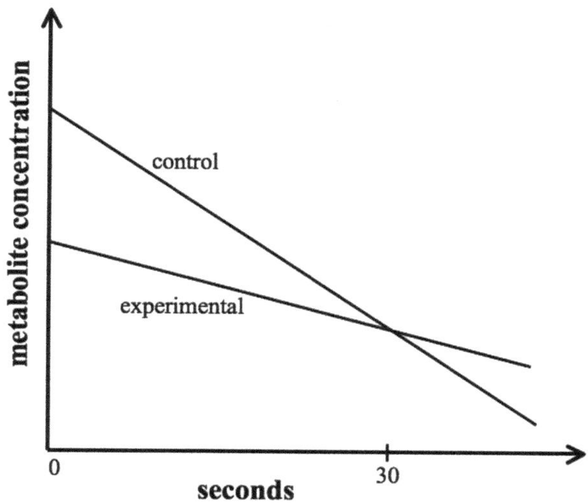

Fig. 1 Figure showing decay of labile metabolites over time unless frozen

freeze blowers, brain choppers, and high intensity microwave ovens were created to minimize killing artifact. These devices yielded data as close to true in vivo values as have ever been seen.

Beriberi

In the past, beriberi was a serious health risk for millions of people. This was largely linked to the consumption of polished rice. Vitamin B1 exists in the outer shells of the rice, which are removed in the process of producing polished rice. Many people were by circumstance forced to eat most meals consisting of almost exclusively polished rice. As overall world diet improved and knowledge increased, the incidence of beriberi dropped.

Nevertheless, outbreaks of beriberi continued to occur. These were often on a smaller scale than before; still hundreds might be affected in these outbreaks. For example, in prison populations a change in rice venders or source could quickly trigger an outbreak in people whose nutritional status was already threatened. If such an outbreak was recognized and treated the outcome could be relatively positive. If medical care was spotty, and patients not treated, then significant sequelae could occur. These outbreaks are frequently under-reported or not reported due to possible political ramifications. This is currently (2009) being witnessed in the cholera epidemic in Zimbabwe.

A solution to the problem of isolated outbreaks of beriberi is actually simple and inexpensive. The first part of a solution must be education. Education about beriberi, how it occurs, how it is treated, etc., should be directed to both healthcare workers and the general population. This can be accomplished in schools, healthcare training, and brochures and fliers widely distributed.

A second aspect of dealing with an outbreak of suspected beriberi is to have some simple test or clinical criteria with which to base a diagnosis. Hospitals in the western world currently use the measure of the thiamine-dependent enzyme transketolase and the effect of added thiamine pyrophosphate on enzyme activity as a measure of thiamine deficiency. This assay in its present form is not a realistic possibility for third-world countries. Some simpler test needs to be available so that healthcare workers can be assured they are pursuing correct treatment procedures.

The third aspect of delivering proper healthcare is to have on hand large numbers of vitamin supplementation including thiamine. The cost of this treatment is almost insignificant relative to the health benefits achieved. The shelf life of such vitamins is measured in years. It should not be a major issue to develop a strategy for distribution to even the most remote areas. It would be easy to develop a distribution plan in which geographic areas with a high risk could be supplied first and regularly. These needs are, of course, not as pressing as fighting AIDS or malaria, but the cost/health benefit is highly positive. World health organizations or grants to local governments (with oversight) should proceed with such plans.

Wernicke's Disease

The impact of alcohol abuse in the US is staggering. It has been estimated that there are at least 15 million alcoholics, and many cases go unreported. Wernicke's disease is thought to occur in about 12% of alcoholics, meaning almost 2 million cases. The cost of this in terms of lost wages, care, etc., is overwhelming. Adding thiamine to alcoholic beverages would go a long way toward reducing the impact of alcoholism and Wernicke's disease. This has been resisted by distillers (for cost reasons), yet folates have been added to food (bread) for decades. Legislation to affect this change should be passed as soon as possible (Figs. 2 and 3).

The lesions – development, location, etc. – of Wernicke's disease are an enigma. All who view them are stricken by the bilateral symmetry and highly localized nature of the lesions. Microscopically, the tissue transforms from normal to diseased in a linear space of one or two cells. The reason for this highly localized transformation is totally speculative. Various reasons which have been advanced include the following: some unique embryological reason, blood flow differences from surrounding tissue, a different energy metabolism in affected tissue, different enzyme reserve capacity, etc.

John Blass and coworkers 30+ years ago showed that various areas of the cerebellum had unique biochemical features. One of these was that the enzymatic abilities as regards pyruvate dehydrogenase, for example, were only marginally greater than flux required, as compared to other brain regions where there was a several fold reserve capacity. These kind of sophisticated assays have not been applied to

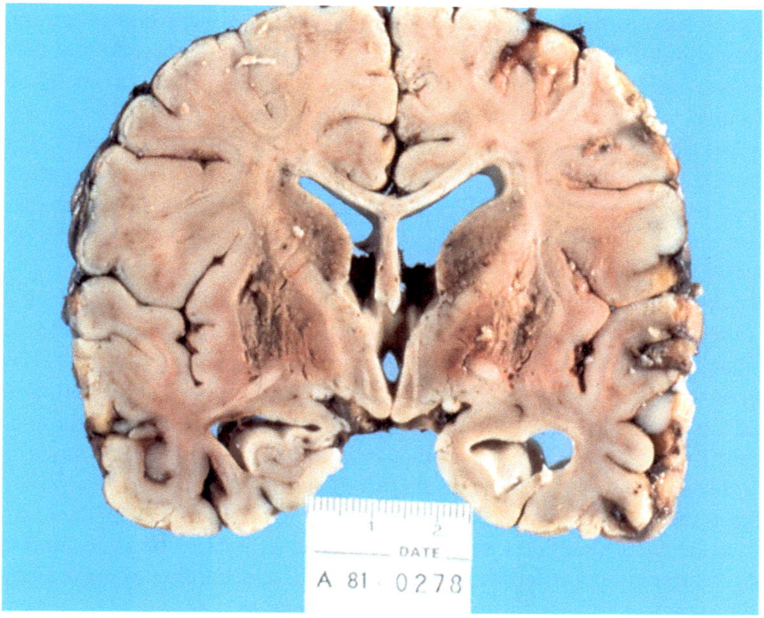

Fig. 2 Brain slice from a Leigh's patient showing basal ganglial lesions, but complete sparing of the mammillary bodies. Figure courtesy of Dr. Michael D. Norenberg

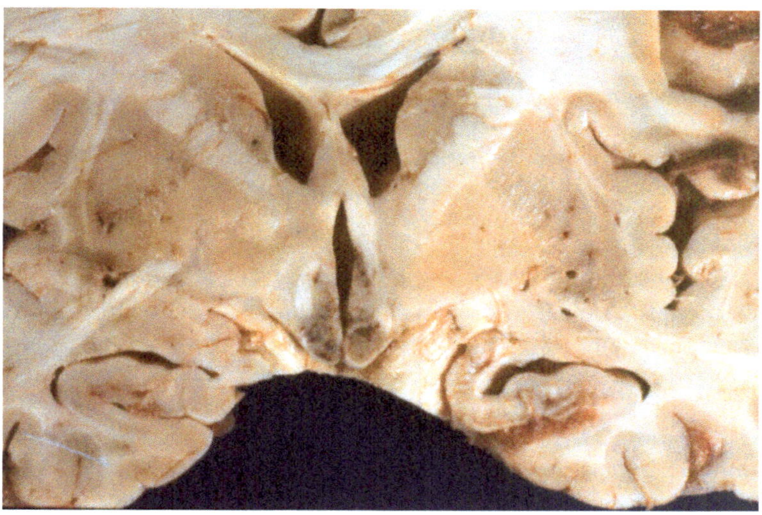

Fig. 3 Brain slice from a patient with Wernicke's disease showing impressive bilaterally symmetrical lesions in the mammillary bodies, with relative sparing above. This figure together with Fig. 2 emphasize the highly selective yet different nature of the lesions in these two related diseases. How is it that such a small anatomic location as the mammillary bodies are severely affected in one disorder, whereas they are not affected in another associated disease? Figure courtesy of Dr. Michael D. Norenberg

small areas such as the mammillary bodies. The mammillary bodies are important in memory/limbic system, and serve as synaptic sites. It is possible that the energy metabolism (ATP) requirements are such that small depletions in metabolic substrates could lead to cell damage and death.

The next comment also relates to the chapter on CPM/ACD/MBD. How is it that chronic alcoholism can lead to the above three plus Wernicke's in patients with largely indiscernible histories? Why is it that some people develop Wernicke's disease and another patient with identical alcohol exposure develops CPM or ACD? Are there genetic reasons? Are there metabolic reasons?

Why do some patients with Wernicke's respond to early and vigorous treatment while others do not? One hypothesis was that if treatment was initiated before structural changes started, then successful treatment was possible. Now, it is recognized that some structural changes can be reversed, especially in patients admitted to the hospital for the first time, implying early stages. This makes early diagnosis and treatment imperative. Some report that many – over half – of all cases of Wernicke's disease go undiagnosed and therefore untreated. This seems a highly unacceptable statistic. Diagnostic and imaging strategies need to be mobilized to assure quick and accurate diagnosis and treatment. The avoidance of great morbidity by missing treatment opportunities far outweighs the added effort to make correct and early diagnoses. Methods for ensuring compliance in diagnosed patients must be stressed.

CPM/ACD/MBD

MBD is a rare consequence to chronic alcoholism, which has the usual highly specific neuropathological lesions. At first it was thought to be a disorder that was found solely in Italian wine drinking males – a highly specific population. Soon however it became apparent that there was no localized population affected by MBD, but that it could affect anyone. Interestingly, there are many reported cases in which early diagnosis and treatment can yield favorable results. The key hope for successful treatment seems to lie in initiating a treatment regime at the earliest time. This would fit the concept that generally speaking, if thiamine therapy can start as early as possible, structural changes can be minimized.

The structural changes that one would hope might not occur are those of neuronyl cell death. Other structural changes such as capillary alterations with extravasation of blood into the lesion area and glial responses may be reversible and may predate unalterable changes in neuron structure and function.

The cerebellar vermis change seen in ACD can neatly be explained, as stated above, by the unique cerebellar metabolism described by Blass and coworkers. The finding of a low reserve capacity of the cerebellum for pyruvate dehydrogenase activity should at least imply that similar circumstances might also underlie changes seen in other very small areas. Thus, lesions in mammillary bodies and thalamic nuclei may have the same underlying biochemical limits as does the cerebellar vermis. Confirmation awaits the properly designed experiments.

One problem with this hypothesis is that of timing and location of lesions. If these lesioned areas of the brain reflect a narrow range of enzyme capacity for catalyzing critical reactions, then how is it that some chronic alcoholics develop CPM while others develop MBD or ACD? Maybe it is "chance" or that there are minor enough differences in enzyme expression to account for the different disease processes. Regarding timing, why do decades pass before biologically significant changes occur in cerebral neurons? Why is there not a "down and out" sequence of events as seen in beriberi outbreaks, the duration of which may occur over a few weeks only? That is certainly the case in ASA in which patients may have been on the edge as regards thiamine-deficient enzymes, then one meal of *Anaphe venata* larvae pushes them into a neurological crisis in just a few hours. It seems unlikely that process exists in Wernicke's cases. Thiamine treatment in cases of ASA produces a rapid response. Thiamine treatment of MBD and Wernicke's disease may also begin reversal of neurological symptoms in just a few hours.

Leigh's Disease

Leigh's disease has had an interesting history. For a number of years, the disease was viewed from a neuropathological viewpoint that regardless of treatment trials which might have resulted in transient improvements, the final outcome was inevitable. Much effort was expended on the urinary inhibitor concept, which was thought to be an "agent" or toxin which if eliminated or rendered inactive might ameliorate the relentless progress of the disease. Unfortunately that was not the case, and it now appears that Leigh's disease consists of many different phenotypic expressions of an equal number of genetic point mutations. Each of these has, in common, alteration of thiamine-deficient enzymatic processes, which in turn produces many, yet, similar case presentations. The likelihood that Leigh's disease ultimately falls under the umbrella of the inherited ataxias is strong.

Thus all Leigh's disease cases may represent mitochondrial defects, for example, in pyruvate dehydrogenase, which lead to very similar phenotypes. But what does this matter if the outcome remains the same?

The correlation is always drawn between lesions in all these cases, especially in the comparison between Wernicke's disease and Leigh's disease. Except for the mammillary body lesions in Wernicke's disease, the lesions are very similar. Thiamine treatment can be beneficial in Wernicke's disease, but except for possible transient improvement in Leigh's disease, thiamine does not work. The transient improvement reported in some cases may have been subjective, or stimulated an increased "blip" in enzyme activity. In the cases of point mutations in genes controlling enzyme function a large amount of thiamine will not change the outcome, but the appearance of the patient will appear as one with thiamine deficiency.

One area which appears not to have been tried is that of enzyme replacement therapy. Dr. William Sly, for one, has successfully treated cases of mucopolysaccharidoses with large doses of enzyme which have crossed the blood brain barrier and achieved improvement in patients with this disorder. This should be tried in Leigh's patients.

Another hypothesis regarding similarities between Leigh's disease and Wernicke's disease relates to the blood brain barrier. Alcohol has a deleterious effect on astrocytes, decreasing their functional capacity. If the blood brain barrier is compromised in alcoholics, then this could actually work in favor of thiamine treatment in that a functionally decreased blood brain barrier would be conducive to the passage of thiamine into the brain. Thiamine in serum is associated with high molecular weight albumen and ordinarily is low in concentration in brain as compared to other tissues. Symptoms of Leigh's disease characteristically start around the second year of life, when the blood brain barrier is functional and could prevent entry of thiamine. Studies of transport of labeled thiamine into brain with increasing age and in the presence and absence of alcohol would answer many questions.

Another area mentioned earlier is that of metabolic rate in lesioned brain sites. Using techniques described by Passonneau and Lowry, metabolite levels and turnover could be measured in vivo in lesioned areas in mice or rats. This would directly answer metabolic questions regarding maturation of energy metabolism as well as effects of thiamine deficiency and alcohol in lesioned areas. Wernicke's disease is a thiamine deficiency problem, and Leigh's disease has strikingly similar brain lesions suggesting similar biochemical similarities.

Inherited Ataxias

The above comments pertain as well to the inherited ataxias. These cases have in common ataxia as a first symptom, and usually at an early childhood age. These all appear to be mitochondrial defects involving several sites, but all having a very similar phenotype. Lactic acidosis, altered enzyme levels (pyruvate decarboxylase) various neurological deficits such as nystagmus, etc. are common features. The ataxia is frequently described as cerebellar ataxia.

Once again, the group of maladies called inherited ataxias represents a collection of defects associated largely with pyruvate dehydrogenase, and sometimes other enzymatic deficits closely connected to thiamine-requiring enzymes. Thiamine treatment has been tried with little success. Ubiquinone treatment is unsuccessful. Friedriech's ataxia, another inherited ataxia has responded to a ketogenic diet, and this treatment regime should be tried in other inherited ataxias using controlled clinical trials.

It seems that there are multiple variations of the inherited ataxias which are similar to those of Leigh's disease. It is important to garner information regarding the exact underlying mechanism of defect in these disorders. Early diagnosis and treatment are essential for ultimate successful treatment regimes. Again, as in

Leigh's disease, enzyme replacement therapy might prove beneficial in some of these disorders.

Serious Illnesses and Thiamine Deficiency

The category of thiamine deficiencies associated with serious human illnesses such as gastric bypass surgery, HIV/AIDS, ALS, and liver failure is one of considerable interest. In some of these cases, such as gastric bypass surgery, the mechanism creating thiamine deficiency is relatively clear. In gastric bypass surgery for tumors, decreased eating might precede surgery. Following surgery, vomiting is a frequent problem in that it further decreases food intake. Studies show a significant occurrence of thiamine deficiency in these patients. This responds rapidly to thiamine treatment. There should be no reason not to routinely administer thiamine to gastric bypass patients.

The anticancer agent 5-fluorouracil is a drug commonly given to cancer patients suffering from a variety of tumors. Many patients receiving 5-fluorouracil show mild signs of thiamine deficiency, or show no signs, but test positive for thiamine deficiency using the transketolase assay and TPP effect. 5-fluorouracil has been shown to diminish the formation of thiamine pyrophosphate from thiamine. A 5-fluorouracil metabolite, fluorocitrate, may act as an inhibitor of the Krebs cycle. How 5-fluorouracil causes impairment of normal thiamine related functions is not clear, but this low level of thiamine deficiency induced by 5-fluorouracil is rapidly reversible with thiamine treatment. That a low level of thiamine deficiency might add to the metabolic burden in cancer patients is clear, so this concomitant problem should be watched for, and treated quickly and aggressively when recognized.

Another related view is that there is an increase in the activity of the pentose phosphate pathway in some cancers in order to produce RNA. Thiamine deficiency in cancer patients might act to decrease flux through the pathway, thereby slowing cancer growth. Until good controlled studies are performed, these questions will remain unanswered.

Thiamine deficiency measured by the TPP effect/transketolase assay in non-alcoholic AIDS patients has shown a nearly 25% occurrence. In autopsy material from these patients, cerebral lesions were similar in location and type to non-AIDS patients dying of Wernicke's disease. While AIDS is a wasting disease, which might be associated with vitamin disease, the exact relation between AIDS and thiamine deficiency remains to be elucidated. Whether thiamine deficiency significantly alters the course of AIDS, treated or untreated, awaits controlled clinical evaluation.

The occurrence of thiamine deficiency in these and other serious illnesses is recognized. Many unanswered questions, and obvious clinical evaluation needs to be done before these questions can be answered. The synergistic interactions of more than one possible life-threatening illness are unknown, but are probably not good. In the very least case, healthcare workers should be aware of this potential for thiamine deficiency, watch patients for signs and symptoms, and initiate therapy as soon as thiamine deficiency is suspected.

Conclusions

It is hoped that this chapter will have highlighted the many unanswered questions related to thiamine deficiency and associated clinical disorders. And perhaps suggested many more. Controlled clinical studies are not always easy to perform, for example, in a pediatric setting, but are essential to provide useful information. In the absence of this type of information, false leads may send investigators down barren pathways. More use of animal models – carefully controlled – should prove beneficial. Experiments can be designed to answer specific questions if there is attention to detail. I am not trying to belabor this point, but what can be gained by asking questions about circumstances in the mammillary bodies, then grinding up whole brain looking for answers? All too frequently investigators use techniques that are "up and running" to look at a question/problem rather than designing methodology to directly answer questions. These circuitous approaches to research are nonproductive. Time is better spent doing it right.

Thiamine deficiency and associated clinical disorders represent an exciting field of study. The earliest observations regarding the reversibility of symptoms in these conditions led to the concept of a biochemical lesion. The reversibility "feature" of thiamine deficiency has proved to be highly beneficial. Symptoms such as opisthotonus and ataxia can be reversed in only hours by thiamine treatment. This can be used to test the importance of biochemical changes in specific areas. Thus, if symptoms revert to normal in 3 hours, while decreases in various metabolites remain unchanged, an experimental biochemical dissection has been performed.

More importantly, the reversal of symptoms carries over to patients suffering from disorders such as beriberi and Wernicke's disease. This type translation is essential. Frequently when recognized early, symptoms of beriberi heart disease, and Wernicke's symptoms such as nystagmus revert toward normal in hours. This feature of biochemical lesions permits direct evaluation of mechanisms in these disorders. This provides hope for enigmatic problems such as seen in Leigh's disease. If an as yet unknown treatment becomes available, results in early diagnosed cases might be apparent in a remarkably short time. This is possible only if attention is paid to detail, and if carefully controlled, and carefully conducted experiments are designed and executed.

A key paper is that of Reynolds and Blass (1976), showing a minimal excess of pyruvate dehydrogenase in the cerebellum. Just as there is a normal range of blood glucose levels, there is likely a normal range for brain enzymes. This may vary from region to region, and from person to person. This could explain why one individual drinking for 25 years develops Wernicke's disease, and another with the same history develops CPM. The different results could be attributable to different ranges of focal brain enzyme capacities. This could awaits experimental verification.

Appendix A
Bontius Translation

A very troublesome sickness which attacks man is called by the inhabitants (of Java) beriberi (which means sheep like). I imagine those with this sickness, with legs raised, and knees shaking, walk and go like sheep. It is a tremor and paralysis which affects the motion and sensations of the hands and feet, and sometimes the whole body.

A cause of this sort of sickness is fat, and dark blood full of phlegm

. . . .*

*Translation past this point is difficult.

D.W. McCandless, *Thiamine Deficiency and Associated Clinical Disorders*,
Contemporary Clinical Neuroscience, DOI 10.1007/978-1-60761-311-4,

Appendix B
Thiamine Synonyms

These are synonyms gleaned from the literature for thiamine.

Aneurin
Aneurine hydrochloride
Anti-beri-beri factor
Anti polyneuritis factor
Antineuritic vitamin
B1
Bewon
Biamine
Bithiamine
Cocarboxylase
Metabolin
Slowten
Thiamine hydrochloride
Thiamine monophosphate
Thiamine diphosphate
Thiamine triphosphate
Thiamol
Thiazolium
Vitamin B
Vitamin B1
Vitamine B1
Vitamine B1 hydrochloride

Appendix C
The Nobel Prize Acceptance Speech of Dr. Christiaan Eijkman, 1929

Antineuritic Vitamin and Beriberi

Beriberi is a disease prevalent, epidemically, in tropical and subtropical regions of Eastern Asia, where rice is the staple food of the natives; it is found elsewhere among sago-eating peoples (Molucca Islands), as well as in South America, in places where rice or cassava meal is the staple diet, as in certain parts of Brazil. However, the disease also occurs sporadically – here and there even with abundant frequency – in the temperate zone and, in some circumstances, in the frigid zone.

The symptoms of the disease are paralysis and numbness starting from the lower limbs, as well as cardiac and respiratory disorders accompanied by dropsy. These latter symptoms rapidly come to the fore, and then we speak of "wet beriberi". Soon the motility and sensitivity disturbances, eventually accompanied by severe muscular atrophy, become more pronounced ("dry beriberi"). Mixed and transitional forms, however, are not rare. Where the paralysis is fairly advanced, the peculiar gait of the patient is noticeable. As the extensors of the foot are paralysed, the patient has to raise the knee up and swing the foot forward in order to avoid stumbling over the downhanging toes.

It is mostly young men in full vigour who are stricken by the acute form of the disease; they not unfrequently die suddenly, in terrible distress through inability to breathe. This fatal issue, even in the more chronic cases, is often due to intercurrent causes, e.g. physical strain, or diseases such as malaria, dysentery, etc.

Although beriberi was described as long ago as in the first half of the 17th century by the Dutch specialist in tropical medicine Bontius, it has only shown itself as a really devastating disease in the Malay Archipelago since about the sixties of the last century, and then generally speaking not so much among the free population as among people living to some extent under constraint, such as soldiers, sailors, prisoners, imported coolies in the mines and plantations, and so forth. It was particularly common in the native prisons; even detainees awaiting trial, sometimes died of the disease. And in the native hospitals it claimed many a victim among people admitted for a relatively minor ailment, such as gonorrhoea or a fractured bone.

During the protracted war – a true guerilla war –which we had to wage with the Achin sultanate, both army and navy suffered very severely from the disease. Newly drafted native troops were often unfit for service after about six weeks and the pitiful remnant had to be evacuated as quickly as possible. For instance, a man might apparently be in good health in the morning, and even gave proof of his skill during target practice, and then fell victim to the disease by evening. An army doctor of that time mentions that at his hospital 18 solider died of beriberi on one day.

European troops also suffered from the disease in Achin, though to a lesser extent than the native troops.

Faced with this serious situation, the Home Government decided in 1886 to send out a commission to investigate the nature of beriberi and its cause. Pekelharing, Professor of Pathology, and Winkler, reader in neurology, both of the Medical Faculty of the University of Utrecht, were in charge, and I was seconded to the commission as an assistant. Winkler immediately established that the disease was essentially a form of polyneuritis (more accurately: multiple neuratrophy), a finding which agreed with results obtained by Balz and von Scheube from investigations conducted by them in Japan. The clinical symptoms were also consistent with this pathological-anatomical diagnosis, which moreover could be confirmed in the living subject by electrical examination (reaction of degeneration, etc.).

At that time there were naturally many different theories as to the cause of beriberi. Two of these were based on the fact that the disease was particularly prevalent among rice eating populations, although – as already mentioned – this is not always the case. On the one hand, rice poisoning was suspected; on the other, a deficiency of the rice diet, but not in the same sense as we now understand it.

With regard to the first theory, the disease was found to occur all too often where the rice was of excellent quality from the culinary point of view. Neither was any actual proof of a preexisting poison in the rice diet found, or even sought. This also applies to other poison theories, e.g. that the disease was due to rotten fish or to "mephitic" gases emanating from the soil.

The second theory was formulated by Van Leent in 1879 in the light of his experience with the East Indies Navy. He considered that an one-sided rice diet resulted in malnutrition owing to its very low content of protein and fat, and that this condition would promote beriberi. He found that the incidence of the disease among native sailors fell considerably when they were put on to a European diet.

There were weighty arguments against this theory, in particular the fact that people with great muscular strength and a well developed panniculus adiposus were often attacked by the disease; and also that European crews, on a diet containing sufficient protein and fat, were not immune, even when they were given virtually no rice at all. In the report made after its return to Holland, the commission rejected this theory because it could not be assumed that malnutrition in itself would result in destruction of peripheral nerves. For this a directly injurious effect on the nerves or their centers would be necessary. In the light of our knowledge today, however, we see that there was a very large grain of truth in Van Leent's theory, and it can scarcely be doubted that a lasting improvement in the health of the native crews could have been achieved if it were possible to introduce the Dutch navy food and persuade the men to take it. It is however a very delicate matter to alter the diet to which people have been accustomed from their youth, and consequently with few exceptions the native sailors soon forsook pea soup with sausage and bacon, potatoes, bread with cheese, etc. and went back to the monotonous and poor rice diet, which they were usually able to procure.

Takaki, who – for reasons similar to those of Van Leent – a few years later changed the diet of the Japanese Navy to the European style, thereby it is reported–achieving lasting success in the fight against beriberi, also had to admit that this measure was difficult to put into effect. "By last years experience," he said, "we have found that most of the men dislike meat as well as bread and we do not know what we shall do next."

However this may be, we failed to check the disease permanently, although at first it did seem to subside a little. But this dimunition does not prove much since in any case beriberi was subject to considerable periodic fluctuations, apart from the fact that the statistics on the disease were not very reliable. For instance it was quite obvious that these statistics frequently included cases of palpitations of the heart, oedema of the feet, as well as new cause of death described as "hydraemia perniciosa tropica" – conditions which should obviously have been ascribed to beriberi. In those days diagnosis of beriberi was clearly taboo.

In the limited time which the commission, for official reasons, had at its disposal it was unable to carry out experimental investigations in every conceivable direction, and it is not surprising that at a time when bacteriology was all-triumphant under the leadership of Pasteur and Koch, the commission should have decided first of all to apply the methods of these two men to the problem in hand. It had all the more reason to take this line since beriberi is a disease with pronounced local and periodic tendencies (as defined by von Pettenkofer). The periodic fluctuations, the prevalence under certain climatic and weather conditions, the epidemic occurrence in certain countries and places, the association with buildings, and finally the cases of transmission of the disease described in the literature – all this pointed to an infection. True, under these circumstances a parasitic disease might have been considered – in particular hookworm, widely spread in the tropics, has been accused of being the main cause of beriberi – but the commission was able to establish that anaemia of any severity is not one of the characteristics of this disease.

I shall not dwell on the results of the bacteriological investigations because they have long since ceased to have any significance, although at that time they were apparently of fundamental importance. Polymorphic bacteria were found in the blood of patients, although not regularly, and degeneration of the nerves could be induced in animals by repeated injections of a coccus culture from them. In its report, following its return to Holland, the commission suggested very tentatively that the causebvof beriberi had thus been discovered. The commission succeeded in isolating a coccus, which it considered identical with the above-mentioned one, from the air in barracks which, from its point of view, could be regarded as infected.

Following the departure of the commission I was entrusted with the task of continuing its investigations, but at first I failed to get any further. The disinfection measures which, as is logical, were recommended had not the hoped-for success. Then a chance happening put me on the right track.

A disease, in many respects strikingly similar to beriberi in man, suddenly broke out in the chicken-house at the laboratory in Batavia, and this called for a thorough study. The symptoms of the disease are as follows: The initial stages, following defective evacuation of the crop some days earlier, are characterized by an unsteady gait. The bird has difficulty in perching and has to exert itself in order not to fall; the legs are spread through weakness, and the knee and ankle joint are bent. The bird frequently collapses and falls over when walking. Finally it remains lying on its side and paresis of the wing muscles now becomes obvious from its vain efforts to get up. Paralysis of the body muscles advances rapidly from below. Within a few days the condition of the bird has so deteriorated that it can no longer eat anything without assistance; although swallowing movements are still produced, the bird is unable to lift its head. Symptoms indicating the onset of paresis of the respiratory muscles then appear. Respiration is slowed down, the beak opens, comb and skin become cyanotic, the neck is bent back and the head drawn in. The bird now becomes more and more soporific, the eyes are covered by the nictitating membrane and the body temperature falls a degree or two centigrade below normal.

This was a case of polyneuritis, as indicated by the symptoms and course of the disease and proved beyond question by a microscopic examination.

With regard to the etiology, our original supposition was not confirmed, i.e. that in the case of the striking epizootic occurrence of the disease an infection was involved. Attempts to induce the infection with material from affected birds or from birds which had died of the disease were inconclusive since ll of the chickens, even those kept separate as controls, were affected. No specific micro-organism or any higher parasite was found.

Then suddenly the disease cleared up and we were unable to continue our Investigations. The affected chickens recovered and there were no new cases. Fortunately suspicion fell on the food, and rightly so, as it very soon turned out.

The laboratory was still housed provisionally and in a very makeshift manner at the military hospital, although it was administered by the civilian authorities. The laboratory keeper – as I afterwards discovered – had for the sake of economy fed the chickens on cooked rice which he had obtained from the hospital kitchen. Then the cook was replaced and his successor refused to allow military rice to be taken for civilian chickens. Thus, the chickens were fed on polished rice from 17th June to 27th November only. And the disease broke out on 10th July and cleared up during the last days of November.

Deliberate feeding experiments were then conducted in order to check more thoroughly whether or not the probable connection between diet and the disease actually existed. It was found for certain that the polyneuritis was due to the diet of cooked rice. The chickens were attacked by the disease after 3–4 weeks, and in many cases somewhat later, whereas the controls which were fed on unpolished rice remained healthy. In many cases, birds suffering from the disease could be cured by a suitable alteration in diet.

This difference between polished and whole rice did not lie in the fact that the former had not kept so well during storage, since cooked rice which had been freshly prepared from the whole kernal would also cause the disease. In addition to this, rough rice, i.e. rice with only the coarse husk removed, which deteriorates much more easily (being attacked by mites, mold fungi, etc.), proved harmless during the feeding experiments. This rice as obtained simply by stamping, still has its inner hull – known as its "silver skin" (pericarpium) – and germ wholly or largely intact. We were then able to conclude from a series of highly varied experiments that the antineuritic principle is situated mainly in these parts of the rice kernal, and indeed of any cereal grain. It can easily be extracted with water or strong alcohol and is dialyzable. I also established that it can be used medicinally either through the mouth or parenterally.

Feeding with other Amylaceae, such as sago and tapioca, had exactly the same results as cooking rice. Since these contain only traces of primary nutrients other than starch, it was impossible to dismiss out of hand the suspicion that the disease was due to simple inanition, especially as it was accompanied by considerable emaciation. And at that time – when the effectiveness of minute doses of the antineuritic principle had not yet been established – the fact that the affected birds recovered when fed on a diet consisting only of meat, could be regarded as being tantamount to the result of an experimentum cruces. On the other hand, by adding meat to the starch-rich diet it was possible to prevent the emaciation, but not outbreaks of the disease, even though these were somewhat delayed. Thus, inanition in itself could not be the main cause of the disease (any more than "protein" or "salt" deficiency), even though it promoted it. I also concluded from this that an antineuritic principle must have been present in the meat, i.e. in the animal organism, but that this principle was gradually used up while the chickens were being fed on a starch diet. And finally the outcome of these experiments was encouraging for me in that one apparent difference as compared with human beriberi had thus been eliminated, for, as already stated, beriberi not infrequently attacks well-fed, strong individuals. Starvation experiments which I conducted at that time also produced a negative result. This was confirmed many times over (Holst, Shiga, and Kusama; Fraser and Stanton, et al.) before I resumed these experiments (in Holland) in view of a communication from Chamberlain et al. (1911) that they had observed in three out of eight cases, polyneuritis among chickens which had been given drinking water without any solid food. This time I obtained a positive result in no less than six out of eight cases. However, it could not be concluded from this alone that the cause of the disease was general inanition, for, apart from the fact that total starvation includes partial starvation, we were able to show that the starved birds which were suffering from the disease, just as birds which had been fed on an unbalanced diet, recovered again, despite continuing loss of weight, when given 8–10 g vitamin-rich yeast per day.

Investigations with other animal species on the East Indies showed that birds such as pigeons and the Indian rice bird can also be used with success. In contrast, results of experiments with mammals were almost entirely negative. This disappointment could, of course, not prevent me from testing whether and to what extent the discovery which had been made as a result of the study of polyneuritis in birds could be used in the fight against human beriberi, although at that time there was still reason enough to doubt the identity of the two diseases. I was therefore accused of not being logical, a reproach which I am happy to pass over without comment.

The obvious thing to do was to test the preventive and curative effect of rough rice in the case of beriberi by means of experiments and controls. Rough rice is the staple diet wherever the rural population in the Indies grows rice for its own use and shelled it by primitive methods. White, or polished, rice is processed mechanically and is one of the blessings of European civilization, of our improved techniques which, for instance, have also changed the colour of bread here in Europe from brown to white. Thought-provoking too was the fact, already referred to, that the free population, which fed itself, was much less subject to beriberi than the population whose freedom was restricted, and which was often dependent on imported and therefore mechanically processed, polished, rice. (As already noted, rough rice is unsuitable for prolonged storage.)

When the projected nutritional experiments on man had just begun, an observation was made by Vorderman which had almost the value of a check experiment. As civilian medical inspector for the island of Java he knew that, in accordance with local custom, the native prisoners were given polished rice in some areas, but rough rice in others. The question, therefore, was to ascertain whether there was a connection between the nature of the staple diet and the incidence of beriberi at the prisons. This on-the-spot investigation was carried out by Vorderman at my request, on behalf of the East Indies Government, and at the same time attention was paid to other factors – apart from diet – with a bearing on the subject. As a result of this research, conducted with admirable proficiency and perseverance, my theory based on the chicken experiments was proved correct. The enquiry covered no less than 101 prisons with almost 300,000 inmates. In brief, the proportion of cases of beriberi in the prisons where polished rice was used as staple diet was some 300 times greater than in those where rough rice was used.

Vorderman's remarkable results did not at once find recognition and were received with scepticism and adverse criticism in some quarters. For me, who had meanwhile been repatriated, the results agreed so unequivocally with those of the chicken experiments that the possibility of coincidence could not seriously be considered. However, the nutritional experiments on which I had already embarked were discontinued, and the alteration in the catering regulations for prisoners, suggested after my departure by Vodermann and Grijns, in the long run did not meet with the necessary cooperation of the authorities. Only when Grijns, who continued my investigations, discovered that a certain bean, the "katjang idju" (Phaseolus radiatus), has the protective and curative action in the case of chickens, successful experiments – with this bean – were made on man (Roelfsema, Hulshoff Pol, and Kiewiet de Jonge). It is, however, obvious that this bean is of far less value as a staple food than a suitable rice diet.

New encouragement then came from British India (Braddon, Ellis, Fletcher, and Fraser & Stanton), on the strength of which in 1910 the following resolution was drawn up by the Far Eastern Association of Tropical Medicine meeting in Manila: "That in the opinion of this Association, sufficient evidence has now been produced in support of the view that beriberi is associated with the continuous consumption of white (polished) rice as the staple article of diet, and that the Association accordingly desires to bring this matter to the notice of the various Governments concerned."

And in Hong Kong, 1912:

> That the accuracy of the opinion of the Association, recorded in 1910, has received further and more complete confirmation by investigators in Japan, China, French Indo-China, the Phillippine Islands, Siam, Netherlands India, the Straits Settlements and The Federated Malay States, namely that beriberi...

Experiments on animals were naturally also used in order to trace the distribution of the antineuritic principle in living organisms. Numerous animal and vegetable foodstuffs which are more or less rich in this principle are listed in the literature on the subject. In practice this is of the greatest importance, since here too it has been found that the vitamin-rich foodstuffs can be used successfully to combat human beriberi. Concentrated extracts of these have also been prepared and these likewise have a pronounced curative action both against polyneuritis in birds and against beriberi.

Until a short while ago, however, no one had succeeded in extracting the active principle reliably and in pure form from these initial materials, although indeed efforts to do so were not wanting. And I was particularly delighted when a few years ago Jansen and Donath were able to report from the laboratory in Batavia – now a splendid improvement on my old workshop – that they had isolated the antineuritic vitamin from rice bran. It is a crystalline substance, obtained in the form of a hydrochloride, an analysis of which gives approximately the formula.

With this substance they conducted experiments in preventive feeding with rice birds and pigeons, and they found that an addition of 2 mg per 1 kg polished rice, i.e. in the ratio of 1:500,000, is sufficient to give protection against polyneuritis.

They sent me approximately 40 mg, sufficient to carry out not only preventive but also curative experiments on pigeons and young cocks. I was able to confirm that such a minute addition gives protection and – which, indeed, could not have been expected otherwise-also has curative power.

Accordingly the human requirement of antineuritic vitamin can be estimated at 1–2 mg per day.

References

Aalsmeer, W., and Wenckebach, K. (1938) Herz und Kreislauf, bei der Beriberi, Berlin

Abarbanel, J. et al. (1987) Neurologic complications after gastric restriction surgery for morbid obesity. Neurology 37:196–200

Absalon, M., Harding, C., Fain, D. et al. (2001) Leigh syndrome in an infant resulting from mitochondrial DNA depletion. Pediat. Neurol. 24:60–63

Adamolekun, B. (1992) Seasonal ataxia syndrome in South West Nigeria. Ethn. Dis. 2: 185–186

Adamolekun, B. (1993) Anaphe Venata entomophagy and seasonal ataxic syndrome in soutwest Nigeria. Lancet 341:629

Adamolekun, B., Adamolekun, W., and Sonibare, A. et al. (1994) A double blind placebo controlled study of the efficacy of thiamine hydrochloride in a seasonal ataxia in Nigerians. Neurology 44:549–551

Adamolekun, B., and Ibikunle, F. (1994) Investigation of an epidemic of seasonal ataxia in Ikare, Western Nigeria. Acta Neurol. Scand. 90:309–311

Adamolekun, B., McCandless, D., and Butterworth, R. (1997) Epidemic of seasonal ataxia in Nigeria following ingestation of thr African silkworm Anaphe Venata: Role of thiamine deficiency. Met. Brain Dis. 12:251–258

Adamolekun, B., and Ndububa, D. (1994) Epidemiology and clinical presentation of a seasonal ataxia in western Nigeria. J. Neurol. Sci. 124:95–98

Adams, R., Victor, M., and Mancall, E. (1959) Central pontine myelinolysis. Arch. Neurol. Psychiat. 81:154–172

Ahoua, L. et al. (2007) Outbreak of beriberi in a prison in Cote d'Ivoire. Food Nutr. Bull. 28: 283–290

Aikawa, H., Watanabe, I., Furuse, T. et al. (1984) Low energy levels in thiamine deficiency. J. Neuropath. Exp. Neurol. 43:276–287

Akbarian, M. (1966) Hemodynamic studies in beriberi heart disease. Am. J. Med. 41:197–212

Albert, J. (1947) Beriberi. Acta Med. Phil. 4:7–21

Aldinger, E. (1965) Effect of thiamine deficiency on potential myocardial contractility. Circ. Res. 16:238–243

Ambrose, M., Bowden, S., and Whelan, G. (2001) Thiamin treatment and working memory function of alcohol-dependent people: Preliminary findings. Alcohol. Clin. Exp. Res. 25:112–116

American Med Assoc-Dept of Foods and Nutrition (1975) Multivitamin preparations for parenteral use. JPEN 3:258–667

Andrews, V. (1912) Philippine J. Sci. 7:39

Arbelaez, A., Pajon, A., and Castillo, M. (2003) Acute Marchiafava Bignami disease: MR findings in two patients. Am. J. Radiol. 24:1955–1957

Arcos, et al. (1964) Effect of thiamine deficiency on the swelling electron transport, and oxidative phosphorylation in rat heart sarcomeres. Biochem. 3:2041–2045

Ashburn, O., and Lowry, J. (1942) Development of cardiac lesions in thiamine deficient rats. Arch. Path. 37:27–33

Ashiru, M. (1988) The food value of the larvae of Anaphe venata Butler (Lepidoptera: Notodontidae). Ecol. Food Nutr. 22:313–320

Attas, M. et al. (1978) Fulminant beriberi heart disease with lactic acidosis: presentation of a case with evaluation of left ventricular function and review of pathophysiologic mechanisms. Circulation 58:566–572

Basu, T., and Dickerson, J. (1976) The thiamine status of early cancer patients with particular reference to those with breast and bronchial carcinomas. Oncology 33:250–252

Basu et al. (1974) The thiamine status of patients with cancer as shown by the rbc transketolase. Int. J. Vit. Nutr Res. 44:53–58

Benit, P., Slama, A., Cartault, F. et al. (2004) Mutant NDUFS3 subunit of mitochondrial complex 1 causes Leigh syndrome. J. Med. Genet. 41:14–17

Bennett, C., Jones, J., and Nelson, J. (1966) The effects of thiamine deficiency on the metabolism of the brain. J. Neurochem. 13:449–459

Berendzen, K. et al. (2006) Therapeutic potential of dichloroacetate for pyruvate dehydrogenase complex deficiency. Mitochondrian 6:126–135

Berkovic, S., Shoubridge, E., Andermann, F. et al. (1991) Clinical spectrum of mitochondrial DNA mutation at base pair 8344. Lancet 338:457–458

Berry, K., and Olszewski, J. (1963) Central pontine myelinosis. Neurology, 13:531–542

Betrosian, A. et al. (2004) Occidental beriberi and sudden death. AMJ Med. Soc. 327:250–252

Blass, J. (1972) Abnormalities in pyruvate dehydrogenase and neurological function. Inter. J. Neurosci. 4:65–69

Blass, J., Avigan, J., and Uhlendorf, B. (1970) A defect in pyruvate decarboxylase in a child with an intermittent movement disorder. J. Clin. Invest. 49:423–432

Blass, J., Cederbaum, S., and Gibson, G. (1975) Clinical and metabolic abnormalities accompanying deficiencies in pyruvate oxidation. In Disorders of energy metabolism. Hommes, F., and van den Berg, C., eds., John Wiley, London

Blass, J., Kark, R., and Engel, W. (1971) Clinical studies of a patient with pyruvate decarboxylase deficiency. Arch. Neurol. 25:449–460

Blass, J., and Lewis, C. (1973) Kenetic properties of the partially purified pyrivate dehydrogenase complex of ox brain. Biochem. J. 131:31–35

Blass, J. et al. (1972) An inherited defect affecting the tricarboxylic acid cycle in a patient with congenital lactic acidosis. J. Clin. Invest. 51:1845–1851

Blass, J. et al. (1976b) Low activities of the pyruvate and oxoglutarate dehydrogenase complexes in five patients with Friedreich's ataxia. New Eng J. Med. 295:62–67

Boniol, S. et al. (2007) Wernicke encephalopathy complicating lymphoma therapy: Case report and literature review. South. Med. J. 100:717–719

Boros, L. et al. (1997) Oxythiamine and dehydroepiandrosterone inhibit the nonoxidative synthesis of ribose and tumor cell proliferation. Cancer Res. 57:4242–4249

Boros, L. et al. (1998) Thiamine supplimentation to cancer patients: A double edged sword. Anticancer Res. 18:595–602

Bray, G. (1928) Avitaminosis. Trans. Roy. Soc. Trop. Med. 22:9–17

Brin, M. (1962) Effects of thiamine deficiency and of oxythiamine on rat tissue transketolase. J. Nutr. 78:179–183

Brin, M. (1967) The oxidation of 14C pyruvate and of 14C ribose in thiamine deficient intact rats. Israel J. Med. Sci. 3:792–799

Brunette, M. et al. (1972) Thiamine responsive lactic acidosis in a patient with deficient low-km pyruvare carboxylase activity in liver. Pediatrics 50:702–711

Burgess, J. A. (1958) Beriberi. Fed Proc. 17:1–53

Butterworth, R., and Heroux, M. (1989) Effect of pyrithiamine treatment and subsequent thiamine rehabilitation on regional cerebral amino acids and thiamine dependent enzymes. J. Neurochem. 52:1079–1084

Butterworth, R. et al. (1991) Thiamine deficiency and Wernicke's encephalopathy in AIDS. Met. Brain Dis. 6:207–212

Butterworth, R., Kril, J., and Harper, C. (1993) Thiamine-dependent enzyme changes in the brains of alcoholics: Relationship to the Wernicke-Korsakoff syndrome. Alcohol Clin. Exp. Res. 17:1084–1086
Calza, L., Manfredi, R., and Chiodo, F. (2005) Hyperlactataemia and lactic acidosis in HIV infected patients receiving antiretroviral therapy. Clin. Nutr. 24:5–15
Celik, Y., and Kaya, M. (2004) Brain SPECT findings in Wernicke's encephalopathy. Neurol. Sci. 25:23–26
Chamberlain, W., et al. (1912) Vitamine. Philippine J. Sci. 7:39–48
Chang, L. (1977) The development of pyruvate carboxylase in rat liver mitochondria. Pediat. Res. 11:6–8
Cheney, D., Gubler, C., and Jaussi, A. (1969) Production of acetylcholine in rat brain following thiamine deprivation and treatment with thiamine antagonists. J. Neurochem. 16: 1283–1291
Clayton, B., Dobbs, R., and Patrick, A. (1967) Leigh's subacute necrotizing encephalopathy. Arch. Dis. Child. 42:467–478
Coakham, H. (1972) Rapid irregular movements of eyes and limbs. Brit. Med. J. 1:45–46
Cole, M., Richardson, E., and Segarra, J. (1964) Central pontine myelinolysis: Further evidence relating the lesion to malnutrition. Neurology 14:165–170
Cooper, J. and Pincus, J. (1967) The role of thiamine in nerve conduction. In Thiamine Deficiency, Wolstenholme, G. E. W., ed., Little Brown Co., Boston, pp. 112–120
Cooper, J., Pincus, J., Itokawa, Y. et al. (1970) Experience with phosphoryl transferase inhibition in subacute necrotizing encephalomyelopathy. New Engl. J. Med. 283:793–796
Coxon, R., and Peters, R. (1950) The effect of oxygen and of cocarboxylase on the formation of citrate and alpha-ketogluarate by pigeon brain homogenates. Biochem. J. 46:300–304
Crome, L. (1970) Subacute necrotizing encephalomyelopathy associated with renal and arterial lesions. Brain 93:709–714
Crosby, T., and Chou, S. (1974) Ragged red fibers in Leigh's disease. Neurology 24:49–54
Cruickshank, A., Tefler, A., and Shenkin, A. (1988) Thiamine deficiency in the critically ill. Intensive Care Med. 14:384–387
De Arce et al. (1966) Pyruvate decarboxylation and its relationship to cardiac myofibrils. Am. J. Physiol. 210:1396–1400
Debray, F. et al. (2007) Long term outcome and clinical spectrum of 73 pediatric patients with mitochondrial diseases. Pediatrics Online, 2006–1866
Debray, F. et al. (2008) Pyruvate dehydrogenase deficiency presenting as intermittent isolated acute ataxia. Neuropediatrics 39:20–23
de Langen, C., and Lichtenstein, A. (1938) Clinical Textbook of Tropical Medicine, Amsterdam
Demaerel, P., and van Paesschen, W. (2004) Marchiafava-Bignami disease. New Eng. J. Med. 351:e10
Denny-Brown, D. (1947) Neurological conditions resulting from prolonged and severe dietary restrictions. Medicine 26:41–113
Desjardins, P., and Butterworth, R. (2005) Role of mitochondrial and oxidative stress in the pathogenesis of selective neuronyl loss in Wernicke's encephalopathy. Mol. Neurobiol. 31:17–25
DeVivo, D., Haymond, M., Obert, B. et al. (1978) Defective activation of the pyruvate dehydrogenase complex in subacute necrotizing encephalomyelopathy (Leigh's disease). Ann. Neurol. 6:483–494
DiMauro, S., Servidei, S., Zeviani, M. et al. (1987) Cytochrome c oxidase deficiency in Leigh syndrome. Ann. Neurol. 22:498–506
Doherty, M., Watson, N., Uchino, K. et al. (2002) Diffusion abnormalities in patients with Wernicke encephalopathy. Neurology 58:655–657
Dondorp, A.M. et al. (2004) Unidentified acids of strong prognostic significance in severe malaria. Crit. Care Med. 32:1683–1688
Doung-ngern, P. et al. (2007) Beriberi outbreak among commercial fisherman, Thailand 2005. SE Asian J. Trop. Med. Pub. Health 38:130–135

Dow, R. (1938) Effect of lesions in the vestibular part of the cerebellum in primates. Arch. Neurol. Psychiat. 40:500–520
Dow, R., and Moruzzi, G. (1958) The Physiology and Pathology of the Cerebellum. University of Minnesota Press, Minneapolis
Dreyfus, P. (1965) The regional distribution of Transketolase in the normal and the thiamine deficient nervous system. Exp. Neurol. 24:119–129
Drummond, J., and Marrian, F. (1927) The physiological role of vitamin B. Biochem. J. 21: 1229–1254
Dunn, H., and Dolman, C. (1969) Necrotizing encephalomyelopathy. Neurology 19:536–550
Durck, H. (1958) Untersuchungen Uben der Pathologische Anatomie der Beriberi, Jena
Eijkman, C., and Vorderman, A. (1897) Verslag. Koninkl. Acad. Wetensch. 6:6–11
Fattal-Valevski, A., et al. (2005) Outbreak of life threatening thiamine deficiency in infants in Israel caused by a defective soy-based formula. Pediatrics 115:233–238
Fehily, L. (1944) Human milk intoxication due to B1 avitaminosis. Brit. Med. J. 3: 590–593
Feigin, I., and Wolf, A. (1954) A disease in infants resembling chronic Wernicke's encephalopathy. J. Pediatr. 45:243–263
Ferracci, F. et al. (1999) Marchiafava-Bignami disease. Arch. Neurol. 56:107–110
Follis, R. (1958a) Beriberi. Fed. Proc. 17:1–56
Follis, R. (1958b) Deficiency Disease, Springfield, Thomas
Foresti, V. et al. (1986) Sindrome da immunodeficienza acquista associata a tubercolosi disseminata ed a encefalopatia cronica di Wernicke. Recent Prog. Med. 77:350–352
Fournier, H., and Butterworth, R. (1990) Effects of thiamine deficiency on thiamine dependent enzymes in regions of the brain of pregnant rats and their offspring. Met. Brain Dis. 5: 77–84
Fukumoto, J., and Suzuki, T. (2007) Marchiafava Bignami disease treated by mianserin hydrochloride in short term evaluated by neuropsychological analysis. Psychogeriatrics 7:25–32
Funk, C. (1911) On the chemical nature of the substance which cures polyneuritis in birds induced by a diet in humans. J. Physiol. 43:395–400
Gamper, E. (1928) Zur Frage der Chronischen Alkoholiker. Deutsche Z Nervenh 102:122–129
Gavrilescu, N., and Peters, R. (1931) Biochemical lesions in vitamin B deficiency. Biochem. J. 25:1397–1409
Geel, S., and Dreyfus, P., (1975) Brain lipid composition of immature thiamine deficient and undernourished rats. L. Neurochem. 24:353–360
Gibson, G. et al. (1988) Reduced activities of thiamine dependent enzymes in the brains and peripheral tissues of patients with Alzheimer's disease. Arch Neurol. 45:836–840
Gibson, G., Jope, R., and Blass, J. (1975) Decreased synthesis of acetyl choline accompany-lng impaired oxidation of pyruvate in rat brain Minces. Biochem. J. 148:17–23
Girard, P., Devic, M., and Garde, A. (1956) L'encephalopathie de Gayet-Wernicke des Alcooliques. Rev. Neurol. 94:493–527
Gold, M., Hauser, R., and Chen, M. (1998) Plasma thiamine deficiency associated with Alzheimers disease but not Parkinson's disease. Met. Brain Dis. 13:43–53
Goswami, P. et al. (2008) MRI findings in marchiafava-Bignami disease with central pontine myelinosis. Ind. J. Radiol. Imaging 16:779–781
Green, R. (1937) Chastek paralysis. Minn. Wildlife Dis. Invest. 3:83–84
Grover, W. D., Green,W. R., and Pileggi, A. J. (1970) Ocular findings in subacute necrotizing encephalomyelitis. Am. J. Ophthalmol. 81:599–605
Grunnet, M. (1969) Changing incidence, distribution, and histopathology of Wernicke's polioencephalopathy. Neurology 19:1135–1139
Gubler, C. (1961) Studies on the physiological functions of thiamine. J. Biol. Chem. 236: 3112–3120
Gubler, C., and Johnson, L. (1967) Enzyme studies in thiamine deficiency. In Thiamine deficiency. Little Brown, Boston
Gudden, H. (1896) Klinische und Anatomische Beitraege zur Kenntniss der Multiplan Alkoholneuritis nebst Bemerkungen Ueber die Regenerations. Arch. Psychiat. 28:643–741

Gurtner, H. (1961) Aneurin und nervenerregung versuche. Helv. Physiol. Pharm. Supp. 11:1–13

Haas, L. et al. (2006) Coma in an alcoholic: Marchiafava-Bignami disease. New Zeal Med. J. 119:1244–1249

Hackel, S. et al. (1953) Effects of thiamine deficiency on myocardial metabolism in intact dogs. Am. Heart J. 46:883–894

Hakim, A. (1986) Effect of thiamine deficiency and its reversal on crerbral blood flow in the rat. J. Cereb. Blood Flow Met. 6:79–85

Hakim, A., and Pappius, H. (1981) The effect of thiamine deficiency on local cerebral glucose utilization. Ann. Neurol. 9:334–339

Harper, H. (1942) Carbohydrate metabolism in thiamine deficiency. J. Biol. Chem. 142: 239–248

Harper, C. (1979) Wernicke's Encephalopathy: A more common disease than realized. J. Neurol. Neurosurg. Psychiat. 42:226–231

Harper, C. (1983) The Incidence of Wernicke's encephalopathy in Australia-a neurolopathological study od 131 cases. J. Neurol. Neurosurg. Psychiat. 46:593–598

Harper, C., Giles, M., and Finlay-Jones, R. (1986) Clinical signs in the Wernicke-Korsakoff complex: A retrospective analysis of 131 cases

Harris, C., Townsend, J., and Baringer, J. (1993) Symptomatic hyponatremia: Can myelinosis be prevented by treatment? J. Neurol. Neurosurg. Psychiat. 56:626–636

Hauschildt, S. et al. (1973) The effect of thiamine deficiency on the contractility and metabolism of isolated rat heart. Int. J. Vit. Nutr. Res. 43:416–425

Hawes, R., Manteiro, E., and Smith, C. (1937) Malayan Br. Med. Assoc. 1:87

Hayashi, T. et al. (2002) Marchiafava-Bignami disease with resolving symmetrical putaminal lesion. J. Neurol. 249:227–228

Henneman, E. (1956) The cerebellum. In Medical Physiology. P. Bard, ed., Mosby Co, St. Louis

Henwood, M. et al. (2006) Reconciling diabetes management and the ketogenic diet in a child with pyruvate dehydrogenase deficiency. J. Child Neurol. 21:436–439

Hess, W., and Akert, K. (1955) Experimental data on role of hypothalamus in mechanism of emotional behavior. Arch. Neurol. Psychiat. 73:127–143

Hess, W., and Messerle, H. (1922) Vitamin B. Z. Physiol. Chem. 119: 176–183

Hirayama, K. et al. (2008) Simultaneously coopererative, but serially antagonistic. Behav. Neurol. 19:137–144

Ho, L. et al. (1986) Deficiency of the pyruvate dehydrogenase compound in pyruvate dehydrogenase complex deficient human fibroblasts. J. Clin. Invest. 78:844–847

Hoefs, S. J. et al. (2008) NDUFA 2 complex 1 mutation leads to Leigh's disease. Am. J. Human Gen. 82:1306–1315

Hommes, F., Berger, R., and Luit-dr-Haan, G. (1973) The effect of thiamine treatment on the activity of Pyruvate dehydrogenase. Pediat. Res. 7:616–619

Hommes, F., Polman, H., and Reerink, J. (1968) Leigh's encephalopathy: An inborn error of gluconeogenesis. Arch. Dis. Child. 43:423–426

Iiizumi, M. et al. (2002) Isolation of a novel gene, CABC1, encoding a mitochondrial protein that is highly homologous to yeast activity of bc1 complex. Cancer Res. 62:1246–1250

Indraccolo, U. et al. (2005) Thiamine deficiency and beriberi features in a patient with hyperemesis gravidarum. Nutrition 21:967–968

Itokawa, Y. (1995) Thiamine and nervous system function: an historical sketch. Met. Brain Dis. 11:1–7

Iwalewa, E., Onayade, O., Oyedapo, O. (2005) Sub-acute toxicity and biochemical effects of Anaphe Venata larvae in mice. African J. Biomed. Res. 8:89–93

Jackson, F., Jackson, R., Delumen, B. et al. (1992) Cassava in Liberia. Ecol. Food Nutr. 28:227–242

Jagadha, V., Deck, J., Halliday, W. et al. (1987) Wernicke's Encephalopathy in Patients on Peritoneal Dialysis or Hemodialysis. Ann. Neurol. 21:78–84

Jansen, B., and Donath, W. (1926) On the isolation of beriberi vitamin. Proc Klin Nederlandse Akad Wetschap 29:1390–1400

Jope, R., and Blass, J. (1975) A comparison of the regulation of pyruvate dehydrogenase in mitochondria from rat brain and liver. Biochem. J. 150:397–403

Jope, R., and Blass, J. (1976) The regulation of pyruvate dehydrogenase in brain in vivo. J. Neurochem. 26:709–714

Jubb, K., Saunders, Z., and Coates, H. (1956) Thiamine deficiency encephalopathy in cats. J. Comp. Neurol. 66:217–227

Juhasz-Pocsine, K. et al. (2007) Neurologic complications of gastric bypass surgery for morbid obesity. Neurology 68:1843–1850

Kamoshita, S., Mizutani, I., and Fukuyama, Y. (1970) Leigh's subacute necrotizing encephalomyelopathy in a child with infantile spasms and hypsarrhythmia. Develop. Med. Child Neurol. 12:430–435

Kant, F. (1932) Die Pseudoencephalitis Wernicke der Alkoholiker Hemorrhagica Superior Acuta. Arch Psychiat Nervenk 98:702–768

Kang, H. et al. (2007) Non specific mitochondrial disease with epilepsy in children. Child. Nerv. Syst. 23:1301–1307

Karapanayiotides, T., Zaloni, I., and Thomaides, T. (2006) Delayed development of Wernicke's encephalopathy after gastrectomy. Arch. Neurol. 63:1026–1027

Karuppagounder, S., and Gibson, G. (2009) Thiamine deficiency: A model of metabolic Encephalopathy ans selective neuronyl vulnerability. In Metabolic Encephalopathy, McCandless, D., ed., Springer, New York

Karuppagounder, S.S. et al. (2008) Thiamine deficiency induces oxidative stress and exacerbates the plaque pathology in Alzheimer's mouse model. Neurobiol Aging. Published online, 11 April 2008

Kashi, M., Henderson, G., and Schenker, S. (2009) Wernicke's encephalopathy. In Metabolic Encephalopathy. McCandless, D., ed., Springer, New York

Katz, J., Landau, B., and Bartsch, G. (1966) The pentose cycle, triose phosphate isomeration, and lipogenesis in rat adipose tissue. J. Biol. Chem. 241:727–740

Ke, Z. et al. (2003) Reversal of thiamine deficiency induced neurodegeneration. J. Neuropath. Exp. Neurol. 62:195–207

Kerr, D., Ho, L., Berlin, C. (1987) Systemic deficiency of the first component of the Pyruvate dehydrogenase complex. Pediatr. Res. 22:312–318

Kim, M. et al. (2007) Acute Marchiafava-Bignami disease with widespread callosal and cortical lesions. J. Korean Med. Sci. 22:908–911

Kinnersley, H., and Peters, R. (1930) Observations upon carbohydrate metabolism in birds. Biochem. J. 24:1126–1136

Koedem, J., and Steyn-Parve, E. (1960) Cardiac metabolism in thiamine deficiency. In Sass-Kortsak, Univ. Toronto Press, Toronto

Koike, M. et al. (1976) Purification and function of alpha-ketoacid dehydrogenases in the mammalian multienzyme complexes. In Thiamine, Gubler, C., Fujiwara, M., and Dreyfus, P., eds., Wiley and Sons, New York

Koike, H. et al. (2001) Postgastrectomy polyneuropathy with thiamine deficiency. J. Neurol. Neurosurg. Psychiat. 71:357–362

Kornreich, L. et al. (2005) Thiamine deficiency in infants: MR findings in the brain. Am. J. Neuroradiol. 26:1668–1674

Korsakoff, S. (1887) Disturbance of psychic function in alcoholic paralysis and its relation to the Disturbance of the Psychic Sphere in Multiple Neuritis of non-Alcoholic Origin. Vestnik Psichiatrii, vol. 4

Korsakoff, S. (1889) A Few Cases of Peculiar Cerebropathy in the Course of Multiple Neuritis. Ejenedelnaja Klinicheskaja Gazeta nos. 5–7

Krampitz, L., and Wooley, D. (1944) The manner of inactivation of thiamine by fish tissue. J. Biol. Chem. 152:9–17

Krill, J. (1995) Neuropathology of thiamine deficiency disorders. Met. Brain Dis. 11:9–17

Krishna, S., Taylor, A., and Supanaranond, W. (1999) Thiamin deficiency complicates malaria in southern Asian adults. Lancet 353:546–549

Krishnan, H. et al. (1945) Beriberi in India. Ind. Med. Gaz. 80:10–23

Kunz, W. (2002) The role of mitochondria in epileptogenesis. Curr. Opin. Neurol. 15:179–184
Kuo, C. et al. (2004) Marchiafava-Bignami disease: CT, MRI, and TC HMPAO-SPECT findings. Chin. J. Radiol. 29:87–91
Lacasse, L., and Lum, C. (2004) Wernicke encephalopathy in a patient with T-cell leukemia and severe malnutrition. Can. J. Neurol. Sci. 31:97–98
Leigh, D. (1951) Subacute Necrotizing Encephalopathy in an Infant. J. Neurol. Neurosurg. Psychiat. 14:216–221
Liang, C. (1962) Studies on experimental thiamine deficiency. Biochem. J. 82:429–434
Lim, E. (1938) J Hong Kong Univ. Med. Soc. 17:215
Liu, Y., Fuh, J., Ling, J. et al. (2006) Correlation of magnetic resonance images with neuropathology in acute Wernicke's encephalopathy. Clin. Neurol. Neurosurg. 108:682–687
Lofland, H. et al. (1963) Enzyme studies in thiamine-deficient pigeons. J. Nutr. 79:188–194
Lohmann, K., and Schuster, P. (1937) Untersuchungen uber die Cocarboxylase. Biochem. Ztschr. 294:188–214
Lowry, O., Passonneau, J., Hasselberger, F. et al. (1964) Effect of ischemia on known substrates and co-factors of the glycolytic pathway in brain. J. Biol. Chem. 239:18–30
Malamud, N., and Skillicorn, S. (1956) Relationship between the Wernicke and the Korsakoff syndrome. Arch Neurol. Psychiat. 76:585–596
Marchiafava, E., and Bignami, A. (1903) Sopra un alterazione del corpo calloso observata in saggertti alcoolisti. Riv. Patol. Nerv. 8:544–549
Mayxay, M. et al. (2007) Thiamine deficiency and uncomplicated falciparum malaria in Laos Trop. Med. Intern. Health. 12:363–369
McCandless, D., and Cassidy, C. (1976) Adenine nucleotides in thiamine deficient rat brain. Res. Comm. Chem. Path. Pharm. 14:579–582
McCandless, D., Curley, A., and Cassidy, C. (1976a) Thiamin deficiency and the pentose phosphate cycle in rats: Intracerebral mechanisms. J. Nutr. 106:1144–1151
McCandless, D., Feussner, G., Lust, W. et al. (1979) Metabolite levels in brain following experimental seizures: The effects of electroshock and phenytoin in cerebellar layers. J. Neurochem. 32:743–754
McCandless, D., and Hodgkin, W. (1977) Subacute necrotizing encephalomyelopathy (Leigh's disease). Pediatrics 60:935–937
McCandless, D., Malone, M., and Szoke, M. (1976b) Pyrithiamine induced thiamine deficiency: Effects on rat myelination. Internat. J. Vit. Nutr. Res. 46: 24–32
McCandless, D., and Schenker, S. (1968) Encephalopathy of thiamine deficiency: Studies of intracerebral mechanisms. J. Clin. Invest. 47:2268–2280
McCandless, D., and Schwartzenburg, F. (1981) The effect of thiamine deficiency on energy metabolism in cells of the lateral vestibular nucleus. Res. Comm. Psychol. Psychiat. Behav. 6:183–190
McCandless, D. et al. (1970) Cardiac metabolism in thiamine deficiency in rats. J. Nutr. 100: 991–1002
McEntee, W., Mair, R., and Langlais, P. (1987) Neurochemical specificity of learning: Dopamine and motor learning. Yale J. Biol. Med. 60:187–193
McKee, A., Winkelman, M., and Banker, B. (1988) Central pontine myelinolysis in severely burned patients. Neurology 38:1211–1217
Mcgowan, G. (1937) Pyruvate oxidation in brain. Biochem. J. 31:1627–1636
Meneegon, P. et al. (2005) Marchiafava-Bignami disease: Diffusion weighted MRI in corpus callosum and cortical lesions. Neurology 65:475–477
Meng, J., and Okeda, R. (2003) Neuropathologicalstudy of the role of mast cells and histamine-positive neurons in selective vulnerability of the thalamus and inferior colliculus in thiamine deficient encephalopathy. Neuropathology 23: 25–35
Mesulam, M., Van Hoesen, G., and Butters, N. (1977) Clinical manifestations of chronic thiamine deficiency in the rhesus monkey. Neurology 27:239–245
Miranda, A., Ishii, S., DiMauro, S. et al. (1989) Cytochrome c oxidase deficiency in Leigh's syndrome. Neurology 39:697–702

Mollet, J. et al. (2008) CABC1 gene mutations cause ubiquinone deficiency with cerebellar ataxia and seizures. Am. J. Human Genet. 82:623–630

Moore, D., Fowler, W., and Crumpler, L. (1990b) 5-fluorouracil neurotoxicity. Gyn. Oncol. 36:152–154

Morcos, Z., Kerns, S., and Shapiro, B. (2004) Wernicke encephalopathy. Arch. Neurol. 61:775–776

Moreadith, R. et al. (1984) Deficiency of the iron sulfer clusters of mitochondrial reduced nicotinamide adenine dinucleotide ubiquinone oxidoreductase in an infant with congenital lactic acidosis. J. Clin. Invest. 74:685–697

Morris, A., Leonard, J., Brown, G. et al. (1996) Deficiency of respiratory chain complex 1 is a common cause of Leigh disease. Ann. Neurol. 40:25–30

Muri, R. et al. (1999) Thiamine deficiency in HIV positive patients. Clin. Nutr. 18:375–378

Murphy, J. (1976) Neurochemical changes in Leigh's disease. Nutr. Sci. Vitaminol. 22:69–73

Murphy, J., and Craig, L. (1975) Leigh's disease: significance of the biochemical changes in brain. J. Neurol. Neurosurg. Psychiat. 38:1100–1103

Murphy, J. et al. (1974) Leigh's disease: Biochemical characteristics of the inhibitor. Arch. Neurol. 31:220–227

Nishimune, T., Watanabe, Y., Okazaki, H., et al. (2000) Thiamin is decomposed due to Anaphe spp. Entomophagy in seasonal Ataxia patients in Nigeria. J. Nutr. 130:125

Nixon, P., Diefenbach, R., and Duggleby, R. (1992) Inhibition of transketolase and pyruvate decarboxylase by omeprazole. Biochem. Pharm. 44:177–179

Nose, Y., Iwashima, A., and Nishino, A. (1976) Thiamine binding protein of *E. coli*. In Thiamine, Gubler, C., Fujiwara, M., and Dreyfus, P., eds., Wiley and Sons, New York

Novak, D., and Victor, M. (1974) The vagus and sympathetic nerves in alcoholic polyneuropathy. Arch. Neurol. 30:273–284

Olson, R., and Piatnek, D. (1951) Conservation of energy in cardiac muscle

Olson, R., and Schwartz, W. (1951) Myocardial metabolism in congestive heart failure. Medicine 30:21–31

Osuntokun, B. (1972) Epidemic ataxia in western Nigeria. Br. Med. J. 2:589

Pannunzio, P. et al. (2000) Thiamine deficiency results in metabolic acidosis and energy failure in cerebellar granule cells: An in vitro model for the study of cell death mechanisms in Wernicke's encephalopathy. J. Neurosci. Res. 62:286–292

Passonneau, J., and Lowry, O. (1993) Enzymatic Analysis: A Practical Guide. Humana Press, Totowa

Patel, M., and Tilghman, S. (1973) Regulation of pyruvate metabolism via pyruvate carboxylase in rat brain mitochondria. Biochem. J. 132:185–192

Phillips, G., Victor, M., Adams, R. et al. (1952) A study of the nutritional defect in Wernicke's syndrome: The effect of a purified diet, thiamine, and other vitamins on the clinical manifestations. J. Clin. Invest. 31:859–871

Pincus, J., Cooper, J., Murphy, J. et al. (1973) Thiamine derivatives in subacute necrotizing encephalomyelopathy. Pediatrics 51:716–722

Pincus, J., Cooper, J., Piros, K. et al. (1974) Specificity of the urine inhibitor test for Leigh's disease. Neurology 24:885–890

Pincus, J., and Grove, I. (1970) Distribution of thiamine phosphate esters in normal and thiamine deficient brain. Exp. Neurol. 28:477–483

Pincus, J., Itokawa, Y., and Cooper, J. (1969) Enzyme inhibiting factor in subacute necrotizing encephalomyelopathy. Neurology 19:841–845

Pincus, J. et al. (1971) Thiamine, TTP, and nervous system lesions in subacute necrotizing encephalopathy. Neurology 21:444–456

Plaitakis, A., Nicklas, W., and Berl, S., (1978) Thiamine deficiency: selective impairment of the cerebellar serotonergic system. Neurology 28:691–698

Platt, B. (1953) Beri-beri. In Handbook of Tropical Dermatology, Simons, R., ed., London, pp. 127–134

Platt, and Alley, (1942) Studies on the metabolism of pyruvic acid in normal and vitamin B1 deficient states. Biochem. J. 33:1525–1537
Pliss, L. et al. (2004) Biochemical and structural brain alterations in female mice with cerebral pyruvate dehydrogenase deficiency. J. Neurochem. 91:1082–1091
Poloni, M. et al. (1982) Thiamine monophosphate in the CSF of patients with amyotrophic lateral sclerosis. Arch. Neurol. 39:507–509
Poloni, M. et al. (1986) Inversion of T/TMP ratio in ALS: a specific finding? Int J. Neurol. Sci. 7:333–335
Rahman, S., et al. (1996) Leigh syndrome: Clinical features and biochemical and DNA abnormalities. Ann. Neurol. 39: 343–351
Ramasatri, B., and Indrauarthi, D. (1957) Beriberi. Ind. J. Med. Res. 45:447–453
Rao, V.L. et al. (1993) Decreased activities of thiamine diphosphatase infrontal and temporal cortex in Alzheimers disease. Brain Res. 631:334–336
Read, D., and Harrington, D. (1986) Experimentally induced thiamine deficiency in Beagle dogs: Pathologic changes of the central nervous system. Am. J. Vet. Res. 47:2281–2289.
Reed, L. (1967) Macromolecular organization of pyruvate and alpha ketoglutarate dehydrogenase compleses. In Thiamine deficiency. Little Brown, Boston
Reed, L. (1976) Regulation of mammalian pyruvate dehydrogenase complex by phosphorylation and dephosphorylation. In Thiamine. Gubler, C., Fujiwara, M., and Dreyfus, P., eds., Wiley and Sons, New York
Reynolds, S., and Blass, J. (1975) Normal levels of acetyl coenzyme A and of acetylcholine in the brains of thiamine deficient rats. J. Neurochem. 24:185–187
Reynolds, S., and Blass, J. (1976) A possible mechanism for selective cerebellar damage in partial PDH deficiency. Neurology 26:625–628
Richter, (1949) The effect of certain quinolone compounds on the nervous system of monkeys. J. Neuropath. Exp. Neurol. 8: 155–170
Richter, R. (1957) Infantile subacute necrotizing encephalopathy with predilection for the brainstem. J. Neuropath. Exp. Neurol. 16:281–307
Rolfe, M. et al. (1993) Urban beri-beri in the Gambia, West Africa. Trans. R. Soc. Trop. Med. Hyg. 87:114–115
Rosenberg, S., Lopes, M., and Tsanaclis, A. (1986) Neuropathology of acquired immunodeficiency syndrome (AIDS). J. Neurol. Sci. 76:187–198
Rosenbaum, M., and Merritt, H. (1939) Korsakoff's syndrome. Arch. Neurol. Psychiat. 41: 978–983
Rosenblum, W., and Feigin, I. (1965) The hemorrhagic component of Wernicke's encephalopathy. Arch. Neurol. 13:627–632
Rydin, H. (1935) The action of crystalline vitamin B1 on the tissue oxidation of chronic avitaminous pigeon's brain. Biochem. J. 29:860–865
Santorelli, F., Shanske, S., Macaya, A. et al. (1993) The mutation at nt8993 of mitochondrial DNA is a common cause of Leigh's disease. Ann. Neurol. 34:827–834
Saudubray, J. et al. (1976) Neonatal congenital lactic acidosis with pyruvate carboxylase deficiency in two siblings. Acta Pediatr. Scand. 65:717–724
Schiff, M., Miné, M., Brivet, M. et al. (2006) Leigh's disease due to a new mutation in the PDHX Gene. Ann. Neurol. 59:709–714
Schramm, C., Wanitschke, R., and Galle, P. (1999) Thiamine for the treatment of nucleoside analogue induced severe lactic acidosis. Eur. J. Anaesthesiol. 16:733–735
Sedel, F. et al. (2008) Therapy insight: inborn errors of metabolism. Nature.com. 3:279–290
Sedel, F., Lyon-Caen, O., Saudubray, J. (2007) Therapy insight: Inborn errors of metabolism in adult neurology. Nature Clin. Prac. Neurol. 3:279–290
Sekiyama, S., Takagi, S., and Kondo, Y. (2005) Peripheral neuropathy due to thiamine deficiency after inappropriate diet and total gastrectomy. Tokai J. Exp. Clin. Med. 30:137–140
Seligmann, H. et al. (1991) Thiamine deficiency in patients with congestive heart failure receiving long term furosemide therapy: A pilot study. Am. J. Med. 91:151–155

Seligmann, H. et al. (2001) Thiamine deficiency in patients with B-chronic lymphocytic leukemia: a pilot study. Postgrad. Med. 77:582–585

Shanske, S. et al. (2008) The G13513A mutation in the ND5 gene of mitochondrial DNA as a common cause of MELAS or Leigh syndrome. Arch. Neurol. 65:368–372

Shen, L. et al. (1968) Interaction between energy charge and metabolite modulation in the regulation of enzymes of amphibolic sequences. Biochemistry 7:4041–4045

Sheu, K., Hu, C., and Utter, M. (1981) Pyruvate dehydrogenase complex activity in normal and deficient fibroblasts. J. Clin. Invest. 67:1463–1471

Shimomura, T. et al. (1998) Development of Wernicke-Korsakoff syndrome after long intervals following gastrectomy. Arch. Neurol. 55:1242–1245

Shimazono, J. (1927) Beriberi. In Avitaminosen und Verwandte Krankheitszustande. Stepp, W., and Gyorgy, P., Ed., Berlin, Springer

Sijens, P. et al. (2008) MR spectroscopy of the brain in Leigh syndrome. Brain Dev. 30:579–583

Simopoulos, A. P. et al. (1972) Subacute necrotizing encephalomyopathy. Neurology 22:1257–1264

Snell, K. (1974) Pyruvate carboxylation in liver mitochondria in the developing rat. Int. J. Biochem. 5:463–471

Steyn-Parve, E. (1967) The mode of action of some thiamine analogues with antivitamin activity. In Thiamine Deficiency, Wolstenholme, G., ed., pp. 26–42

Stoltz, E., and Bessey, O. (1942) The blood lactate-pyruvate relation and its use in experimental thiamine deficiency in pigeons. J. Biol. Chem. 142:625–631

Suzuki, T. (1967) Electron microscopic study on myocardial lesions in thiamine deficient rats. Tohoku J. exp. Med. 91:249–259

Swank, R. (1940) Avian thiamine deficiency. J. Exp. Med. 71:683–702

Takahashi, K., and Nakamurth, H. (1976) Axonyl degeneration in beriberi neuropathology. Arch. Neurol. 33:836–841

Tattevin, P. et al. (2006) Confusion, memory disorders, and ophthalmoparesis in a patient with AIDS. Lancet 367:368

Taylor, S., Mukherjee, C., and Jungas, R. (1975) Regulation of pyruvate dehydrogenase in isolated rat liver mitochondria. J. Biol. Chem. 250:2028–2035

Tobita, M. et al. (1997) A case of Marchiafava-Bignami disease with complete recovery. Tohoku J. Exp. Med. 182:175–179

Tomita, I. et al. (1974) Coenzyme analog inhibition in the reconstitution of yeast pyruvate decarboxylase and transketolase. Biochem. Biophys. Res. Comm. 57: 78–84

Tomlinson, B., Pierides, A., and Bradley, W. (1976) Central pontine myelinolysis: Two cases with associate electrolyte disturbances. Q. J. Med. 45:373–386

Toshima, K., Kuroda, Y., Hashimoto, T. et al. (1982) Enzymologic studies and therapy of Leigh's disease associated with pyruvate decarboxylase deficiency. Pediat. Res. 16:430–435

Tran, H. (2006) A 74 year old woman with increasing dyspnea. Arch. Pathol. Lab. Med. 130:8–10

Tsao, C., Herman, G., Boue, D. et al. (2003) Inherited ataxias. J Child Neurol. 18:62–64

Valanne, L. et al. (1998) Neuroradiologic findings in children with mitochondrial disorders. Am. J. Neuroradiol. 19:369–377

Valyasevi, A., and Vimokesant, S. (1968) Chemical composition of breast milk in different locations of Thailand. J. Med. Assoc. Thailand 51:348–353

van Zaanen, H., and van der Lelie, J. (1992) Thiamine deficiency in hematologic malignant tumors. Cancer 69:1710–1713

Vazquez, C. et al. (2008) Marchiafava-Bignami disease in Uruguay. Neurologia 23:322–328

Victor, M., and Adams, R. (1961) On the etiology of the alcoholic neurologic diseases with special reference to the role of nutrition. Am. J. Clin. Nutr. 9:379–397

Victor, M., Adams, R., and Collins, G. (1989) The Wernicke Korsakoff Syndrome. F.A. Davis Co., Philadelphia

Victor, M., Adams, R., and Mancall, E. (1959) A restricted form of cerebellar cortical degeneration occurring in alcoholic patients. Arch Neurol. 1:579–688

Vorhees, C., Schmidt, D., Barrett, R. et al. (1977) Effects of thiamine deficiency on acetylcholine levels and utilization in rat brain. J. Nutr. 107:1902–1908
Walker, U. (2004) Lactic acidosis in HIV patients – diagnosis and treatment. MMW Fort. Med. 146:65–67
Wallis, W., Willoughby, E., and Baker, P. (1978) Coma in the Wernicke-Korsakoff syndrome. Lancet 2:400–401
Wang, X. et al. (2007) Thiamine deficiency induces endoplasmic reticulum stress in neurons. Neuroscience 144:1045–1056
Watanabe, I., Tomita, T., and Hung, K. et al. (1981) Edematous necrosis in thiamine deficient encephalopathy of the mouse. J. Neuropath. Exp. Neurol. 40:454–471
Weil-Malherbe, H. (1940) Thiamine deficiency. Nature 145:106–107
Wernicke, C. (1881) Lehrbuch der Gehirnkrankheiten fur Aerzte und Studirende. Vol. 2 Kassel, Theodor Fischer, pp. 229–242
Williams, R. (June 1936) Structure of vitamin B1. Amer J. Chem Soc. 58:1222–1225
Williams, R., and Cline, J. (Aug 1936) Synthesis of vitamin B1. Amer. J Chem. Soc. 58:1504–1505
Wright, F., and Morley, D. (1958) Encephalitis tremans. The Lancet i:871–873
Yashon, D., and Jane, J. (1967) Subacute necrotizing encephalomyelopathy of infancy and childhood. J. Clin. Path. 20:28–37
Yoshinori, I. (1995) Thiamine and nervous function: An historical sketch. Metab. Brain Dis. 11:1–7
Yoshitoshi, B. et al. (1961) Experimental studies on the beriberi heart. Jpn. Heart J. 2: 42–64
Zieve, L., Doizaki, W., and Stenroos, L. (1968) Effect of magnesium defeciency on blood and liver transketolase activity and on the recovery od enzyme activity in thiamine-deficient rats receiving thiamine. J. Lab. Clin. Med. 72:268–277
Zuccoli, G. et al. (2007) Wernicke encephalopathy: MR findings at clinical presentation in twenty six alcoholic and non alcoholic patients. AJNR Am. J. Neurol. 28:1328–1331

Index

Note: Appendix pages have been indexed with appendix number in brackets followed by "app"

MIX
Papier aus verantwortungsvollen Quellen
Paper from responsible sources
FSC® C105338

If you have any concerns about our products,
you can contact us on
ProductSafety@springernature.com

In case Publisher is established outside the EU,
the EU authorized representative is:
Springer Nature Customer Service Center GmbH
Europaplatz 3, 69115 Heidelberg, Germany

Printed by Libri Plureos GmbH
in Hamburg, Germany